S'energy and The Concept of the Prime Wave

Robert D. Shriner

S'energy and The Concept of the Prime Wave

S'energy
and
The Concept of the Prime Wave

Reality Physics

The Origin of Time.

The Origin of π.

The Origin of the "God Particle".

The Origin of Physical life in a living Universe.

An original concept by

Robert D. Shriner

S'energy and The Concept of the Prime Wave

I dedicate this book to my family for putting up with me and my dreams.

All rights reserved.
Copyright © 2012 by Robert D. Shriner
ISBN-13: 978-1482787221
ISBN-10: 1482787229

S'energy and The Concept of the Prime Wave

CONTENTS

		Page:
FORWARD		7
CHAPTER ONE	ABSTRACT VS. PHYSICAL	21
CHAPTER TWO	"SOMETHING IS MISSING"	27
CHAPTER THREE	ONE STEP UP AT A TIME	43
CHAPTER FOUR	PONDER A MOMENT	53
CHAPTER FIVE	LEVEL ONE - S'ENERGY	57
CHAPTER SIX	LEVEL TWO - PRIME WAVE	61
CHAPTER SEVEN	LEVEL THREE – PACKET	69
CHAPTER EIGHT	LEVEL FOUR – PARTICLE	79
CHAPTER NINE	LEVEL FIVE – ATOM	91
CHAPTER TEN	LEVEL SIX – MOLECULE	99
CHAPTER ELEVEN	LEVEL SEVEN – COSMOLOGIC	105
CHAPTER TWELVE	WHAT IF	111
CHAPTER THIRTEEN	"SOMETHING IS MISSING" REVISITED	123
CHAPTER FOURTEEN	MINDS EYE	137
CHAPTER FIFTEEN	SIMPLE	145
CHAPTER SIXTEEN	POINTING THE WAY TO RECONCILLIATION	153
IN CONCLUSION		163
REFERENCES		167
TABLE OF FIGURES		169

S'energy and The Concept of the Prime Wave

S'energy and The Concept of the Prime Wave

FORWARD

"An idea believed to have value left hidden in the mind is far worse than never having had the idea in the first place for it becomes a complete waste of ones limited and valuable time."
(R. Shriner 1973)

Forty years ago, and from time to time as now, I wrote of Substantive energy, or S'energy, and the concept of the Prime Wave, a comparatively simple thought of seven, simple to complex, levels of definition of a single process called the physical.

- Consider a thought that the descriptions of the physical, in current theory, always begins from the top down and, even with the successes that we have witnessed, this view has taken us to a perceptual point to consider 8 additional dimensional space bound to pure, abstract, self-isolating formulations that today have no known means of being tested in our 3 dimensional space.

- Constants of Nature exist and have value yet none have a foundation in theory for that value to be what it is.

S'energy and The Concept of the Prime Wave

- Cornerstone concepts and theories of the current thinking on physics have not been reconciled.

- Fundamental Forces of Nature exist as conditions but conditions that have no source, nor means, and do not have a known mechanism for an action or reaction to occur.

- Time has been successfully elevated to a status of a fourth physical dimension and there is no known reason or means for the physical to change in a consistent, comparative manner in only 3 dimensions.

- What is the "Higgs' Boson" or the "god particle" that eludes our greatest efforts to explain?

- What is the basis for Dark Matter?

- And what is the basis for Dark Energy?

These are common unanswered questions beyond the questions posed over a lifetime but to this list, formidable as it is, I would like to add two more.

- What is the foundation for Pi?

- And is there a basis for Physical life?

S'energy and The Concept of the Prime Wave

Imagine, if you will, the idea that there are perhaps as many "theories" about how it all fits together as there are people who exist and who have ever been. The world of the scientist is one that must deal with the dreamers and the scientific cranks and the zealots of the day and they have developed a mechanism for dealing with this problem...ignore them. If you do not have the training or the ability to gather to you the knowledge and the methodology of the past and present them then it is quite possible and highly probable that you have made a mistake with your great idea and have rationalized the greatest obstacle to insignificance or taken a major leap of faith to begin where no one has been and no one else cares to go. I do not represent myself as having the training necessary to present this concept for peer review. I do not have the ability to expand my thinking into the world of the mathematicians mind, yet, again, I have a belief that an idea believed to have value left unstated is far worse than never having had the idea at all for it would be a complete waste of time. There are no proofs in this presentation. There is no math in this presentation. But there is a logic and a desire to understand and I have taken the time to work on an idea

S'energy and The Concept of the Prime Wave

that I believe has value. My quest was to understand the nature of Nature, to understand how it all fits together. I sought out the teachings of the scientists of the day and, quite simply stated, they have not satisfied my needs. It is my hope that others might read and understand what I seek and what I seem to have found in my journey.

"So it seems to me to be as I now see the thing I think I see"
Anonymous

"The What is it question transitioned into a What if question that began a personal journey of a lifetime." (Shriner 1973)

Early on, in a dusty old handbook of physics under the title Theoretical Physics, I was taken aback by the statement, if my memory serves me right, of <u>do not presume to ask the what is it question rather be satisfied with the what it is like answers</u>. To this day I find it hard to accept such advice on a subject that is quite possibly one of the most asked questions posed by man... How does it all fit together? As a lifetime student of the sciences, it would be a rare exception if I ever passed on any news of the new achievements and advancements and studies in such magazines as Nature and Scientific American or specials on the TV or radio and certainly from any books by or about Einstein,

S'energy and The Concept of the Prime Wave

Hawking, or Weinberg or Russell or Gamow, or Sagan, or Asimov, or Wheeler, or Newton, just to name a few. I have witnessed an evolving process including Star stuff, Black holes, and Virtual particles with all of their color and charm. Through it all, I found it quite natural to visualize in three dimensions the relationships such as using a bowling ball on a rubber sheet as a depiction of the condition of a gravity field around a body with mass. I continued looking while knowing that the problem is three dimensional and what is taught is visualized more like a light in the fog. Through it all, I also became very much aware of the dependence upon pure, abstract, self-isolating, successful in prediction, mathematical formulations that may or may not have a known physical counterpart. It seems that for mathematical science it was enough to predict an outcome and not get bogged down by what was actually happening. This situation was emphasized by Dr. Wheeler when he spoke of the successes of mathematics and physics and still "...something is missing." The something that is missing, was to me, a clearly understood physical process...but few where looking in that direction after Dr. Einstein introduced an abstraction called

S'energy and The Concept of the Prime Wave

Time as a fourth physical dimension. Dr. Einstein successfully denied a three dimensional physical means or a perceptual Euclidian space, for time and for gravity as a Fundamental Force for a force required an "action at a distance" while a field, a gravitational field, was simply the metric properties of the space-time continuum. An object with mass would simply fall toward another object with mass in a predictable way. And further even the notion of falling is removed as there is no up or down only distance, a direction, a probability of existence, and symbols that represent possible conditions. What an enigma to be able to predict a physical process using pure abstractions that have no known physical counterpart and <u>we call it physics</u>.

Dr. Einstein imposed a great inspirational but intuitive description to a portion of a physical process that certainly was not wholly understood then or now. For a hundred years we have lived with the idea that the micro and the macro worlds we perceive to exist do not need to match, in a mechanistic way, what our senses teach us every day of our lives. It is acceptable to be able to predict an outcome no matter what caused that outcome to occur. Yet and still *"...something is missing!"*.

S'energy and The Concept of the Prime Wave

So... what if there where a single process that underlies, and forms a foundation, that builds up the ladder of complexity from the most fundamental condition to arrive at an ability to write about writing about it, or to perceive the perception?

So... what if there where a means for time, for consistent, comparable, physical change to occur?

Lincoln Barnet quoted Einstein in his 1948 book "The Universe and Dr. Einstein" *A Bantam Book, 1948,a thought* concerning a complete Unified Field Theory is:

"to cover the greatest number of empirical facts by logical deduction from the smallest possible number of hypotheses or axioms."

and Mr. Barnett went on to say that:

"The urge to consolidate premises, to unify the concepts, to penetrate the variety and particularity of the manifest world to the undifferentiated unity that lies beyond is not only the leaven of science; it is the loftiest passion of the human intellect. The Philosopher and the mystic, as well as the scientist, have always sought to arrive at knowledge of the ultimate immutable essence that under girds the mutable illusory world."

What if there where a lowest common denominator for all things physical? What would its qualities have to be?

S'energy and The Concept of the Prime Wave

We commonly use the term energy, an abstract term that can be thought of as meaning "all not otherwise defined" and to represent a potential to do work but what if energy has a physical counterpart also meaning "all not otherwise defined"? This energy, this lowest common denominator, must exist in a volume of space and must be dynamic and must have fundamental qualities that support consistent and comparable change, or time while also being capable of supporting the development of a structure with a geometry that is also consistent and comparable. Supportive of this existence is from I. Newton who represented that <u>*energy cannot be created nor destroyed only modified*</u>. This substance, here in called Substantive energy, or S'energy, fulfills the requirement for the *ultimate immutable essence* in its existence as the lowest common denominator. For S'energy to exist and be dynamic would fulfill the requirement for *under girding the mutable*. The proposal that there is only dynamic S'energy would fulfill the requirement of *"to cover the greatest number of empirical facts by logical deduction from the smallest number of hypotheses or axioms."* Building from the bottom up by describing the qualities of S'energy and exposing the process

S'energy and The Concept of the Prime Wave

by which consistent comparable change can occur and then, level upon level, demonstrate the mechanism of structure, of forces and fields. Such an embodiment of S'energy becomes a foundation and a process that when viewed from one perspective supports the conclusions of Relativity, or when viewed from another perspective supports the conditions of the Quantum. Space-time can be thought of as a **net** affect condition and the quantum can be thought of as a **specific** affect condition and both are believed to be as a result of Prime Waves of S'energy in S'energy. The Photon with both wave and particle duality is true, as represented in this concept of the Prime Wave, as the Photon is the Packet of level three consisting of sets of individual Prime Waves interacting in the harmonic to form a Packet of sets and is the first stable structure which is the foundation for level four, the Particle. Time begins and begins with each new Prime Wave that in sets progress and converge to form the conditions of Structure, of Forces, of Fields, of Charge, and of a Geometry of interactive harmonics at each level of definition. As you explore the existing and common definitions of the Molecule, the Atom, the Particle, the Quark, you are accepting the notion that

S'energy and The Concept of the Prime Wave

by definition and for ease of discussion the summary terms represent a whole host of bits and pieces that would be quite cumbersome to have to repeat and so too is the development of the concept of the Prime Wave tracking upward through simple interactions being defined at one level and the complex of that same level of definition becoming the simple aspect of the next level. I mention this here as it is common to think in top down terms and it is uncommon to think in bottom up terms and to merge these two opposed views is, as it turns out, to be quite a problem. An interpretive problem exists not so much in the development of the structures but more a problem in the presentation whereby a preconceived notion of a relationship maybe imposed on the bottom up developments prematurely. Working from the top down there is no known mechanism for consistent comparable change called time yet working from the bottom up, physical change, or time, begins and begins at the beginning of level two in the definition of the Prime Wave. Please to understand that there are hundreds if not thousands of unanswered questions in the applications of the existing Cornerstone concepts and theories and to address each and every

S'energy and The Concept of the Prime Wave

one in a single presentation would be counterproductive to the presentation and with this being said I will still try to expose the conditions developed in the concept of the Prime Wave that will lead to a resolution of those unanswered questions. As an example I write of the structure of a Particle at level four and show that there are six zones of interaction. These zones of interaction are as a result of the development of the lower three simple to complex levels of definition and so are not ignored. Zone one is the core of the particle and it consists of sets of Prime Waves progressing inward and with some merging into dependent high density nodes that achieve a threshold density and become a source of new Prime Waves. These dependently developed new Prime Waves created in the core of the Particle is a <u>Higgs Particle</u> that is specific to the geometry of the Particle itself. These new Prime Waves act as a time delay of throughput mechanism that seeks to return the Particle back to where it was and does so in a constantly updated time frame associated with the time it takes a Prime Wave to travel to the core of the Particle and back to its source at zone two. If you break the particle geometry then the dependent nodes of the Higgs Particle no

S'energy and The Concept of the Prime Wave

longer occurs and thus the Higgs Particle no longer exists. In the preceding sentences I have focused on the condition of the Higgs Particle but in doing so I have also exposed the conditions of geometry of a Particle with specific interactive zones and in so doing I am addressing the idea of a basis in theory for the Constants of Nature. In addressing the interactive of the Prime Waves in the core of a particle I am also addressing the geometry and the condition of Prime Waves being created and progressing inward and outward from the shell of zone two and that zone three is a condition of these same Prime Waves, that make up the condition of the Higgs Particle on the inward path, forming a specific convergence pattern of the charge field on the outward path that again is associated directly with the geometry of the Particle. Thus a particle with the same geometry of interactive zones will have the same mass and charge and stability, through all variations, as any other with the same geometry no matter where or when.

Such a structure as the Particle with six zones does not exist except by summary definitions in level four as defined and supported by all that precedes and becomes part of all that

S'energy and The Concept of the Prime Wave

proceeds from this or any point of reference. Take a two dimensional slice through a Particle and all you will see is Prime Waves being created, progressing, and converging in a local area thus it is the time element of the harmonics that brings the condition called a Particle into being as is the case for all structures.

This concept that S'energy exists in a volume of space and changes in a consistent and comparable way was my conclusion that I present to you now.

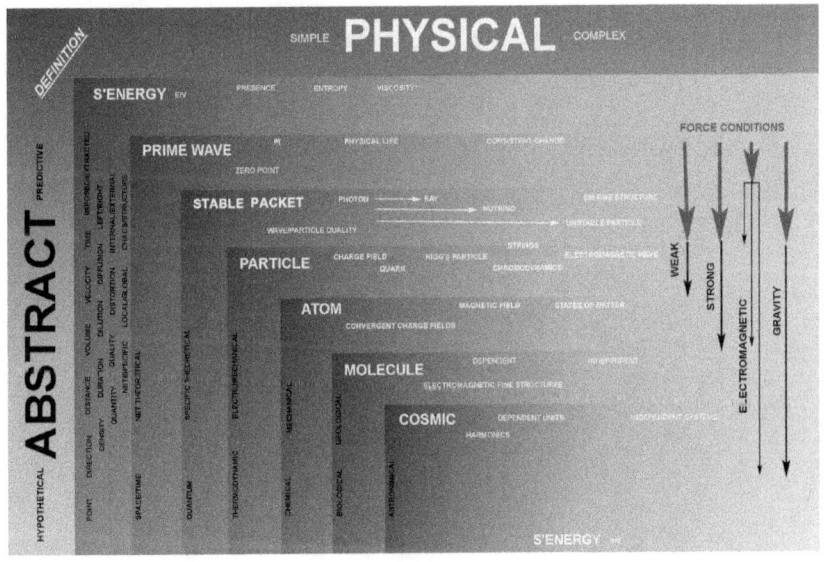

Figure 1

ABSTRACT VS. PHYSICAL PROCESS OVERVIEW.
The view from 30,000 ft. is to see that one level merges into the next higher level and is part of the next lower level of a defined single physical process of the action of S'energy within S'energy.

S'energy and The Concept of the Prime Wave

CHAPTER ONE

The Abstract vs. the Physical

There is one point that needs to be addressed before I continue - the Abstract vs. the Physical. First, I think, we can all agree that 1 + 1 = 2. In this formula rests an unspoken understanding that the one and the other one are equivalent and this grants some certainty to the idea that you have two something's. So long as you do not begin to identify what the one object is and further to identify what the other object is then we all can be comfortable. But in the world of Physics and the mathematics that support it, we truly do not know what the first object is and we do not know what the second object is but we have two of them and we can predict a far more complicated process, that we do not understand, using mathematics again without knowing what the objects are. So what is real and what is an abstract. The objects

S'energy and The Concept of the Prime Wave

are real, in this argument, and the symbols of 1 and + and 1 and = and 2 all have meaning but again it is understood that the number 1 represents a real object, or a defined something, that can be quantified and represented as 1 whole something. I have one nail and I have one house and each of these objects are whole units of one so I can say that I have two whole units. The first is made of 100% steel the second is made of ~100% wood so it can be said, in speaking about the properties of the two objects combined, that you have two objects that are statistically 50% steel and 50% wood. Also one object you can live in and one object you cannot and so on. You now have the potential for a nail to be ~50% wood that you can live in ~50% of the time. I am not building a mathematical proof here but I am pointing out that the use of Mathematics in today's physics has granted some great leeway with insights and predictions. I remind the reader to be aware of the potential for error in the representation of an object only in abstract terms that have an ability to be put together in a specific way so that the predicted outcome is similar to what would be found in an experiment and if not <u>the formula</u> is adjusted or the test is designed with a specific conclusion

S'energy and The Concept of the Prime Wave

supported. If a formulation can predict an outcome then that formulation must be right or correct, <u>in some way</u>, but what about the objects that the formulation represents. This is an extremely small and insignificant example of a huge debate that has been going on in the world of Physics for over a century. Again I mention Dr. J. A. Wheeler formerly of Princeton Univ., in speaking on this subject of the successes of mathematics and physics where he ended with a simple statement that with all of the success still "...something is missing." String Theory, touted as being the future of Physics, allows for three physical dimensions of X and Y and Z and then adds eight more abstract dimensions for a total, today, of eleven dimensional physics and the use of these dimensions seems to be able to connect the dots. This poly-dimensional physics exists today only in the abstract as there is no known means for or to test this thinking in a real world experiment but is this the future of physics? Again I repeat what Dr. Wheeler said "...something is missing." Without a known physical or mechanical means for consistent, comparative, physical change called time - a formulation was developed that could predict and in that prediction stood a

S'energy and The Concept of the Prime Wave

problem with a definition of the object or lack of an object to support the successful prediction so it became acceptable to define time as an abstract physical dimension or condition that, in space-time, modified the space around an object with mass <u>without a mechanical means to do so</u>. In fact any reference to a physical means for time became an act of stupidity for *there is none*. Yet and still, to me, "...something is missing" when you define the physical in pure self-isolating, accurate in prediction abstract terms. To me the physical exists and changes in a consistent and comparable manner and does so without definition while the abstract exists in the mind and needs a definition in that existence...beware of the definition applied to an abstract term that defines a physical process when we do not know what that physical process is. It is this physical process that needs to be defined. It is this physical process that is not understood. It is a clear understanding of the physical process that is the "...something" that is "missing" and both an abstract system and a perceived physical process are needed to approach anything near a Theory Of Everything, or a Grand Unification, or an understanding that is predictable and also useful. Einstein was

S'energy and The Concept of the Prime Wave

able to predict a physical process using an abstract term called time and did so without a known physical means for time to exist, as a condition of consistent comparable change. Einstein developed, in a brilliant intuitive leap, the idea that there is not and cannot be a three dimensional physical or mechanical means for such change to occur. To this last part I disagree with him in the denial of a mechanical means.

Within the concept of the Prime Wave there is a level by level development of a dynamic physical process that begins with substantive energy, or S'energy, existing in an undefined volume of space. (Space in this context is an undefined volume that is unable to change of and by itself.) At level one, where S'energy is defined as having three fundamental qualities, there is change occurring but no specific, consistent and comparable change occurring within level one and so time in the abstract cannot be applied, if you wish, it is not until a mechanism occurs of and through the Prime Wave of level two that a defined consistent and comparable means for change of the physical is identified. In the early years of the 19th century a Particle was nothing but an inert gob of matter? floating in a passive ethereal void that

S'energy and The Concept of the Prime Wave

was shaped by a fourth dimensional thing called time. In the concept of the Prime Wave three additional levels of interaction and defined structure exist below the level of the Particle at level four, and in one of them, level two, is the defined means for consistent comparable change of the physical process called time.

CHAPTER TWO

"SOMETHING IS MISSING"

What we have today:

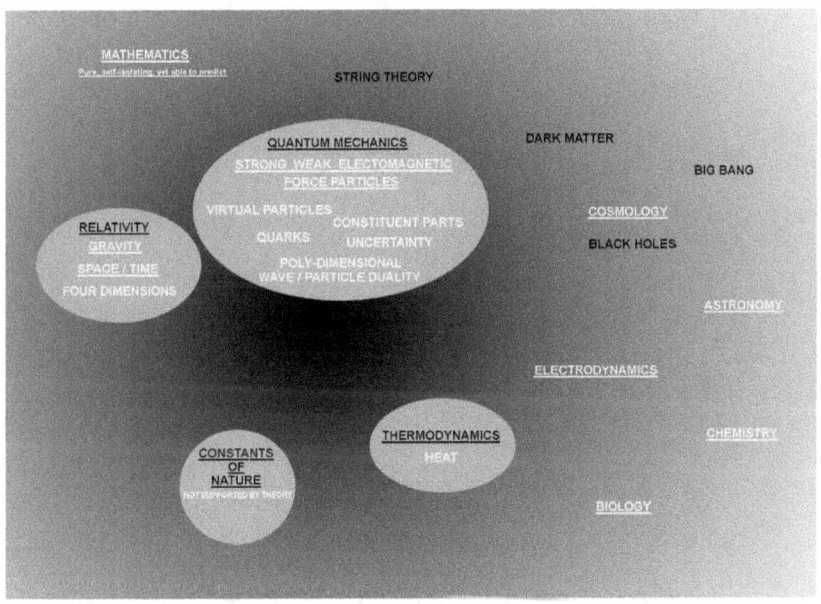

Figure 2

CURRENT PHYSICS 2012

Some of the cornerstone concepts of today that are not reconciled and exist without a foundation that will tie them all together

S'energy and The Concept of the Prime Wave

- Cornerstone concepts and theories have not been reconciled.
- Fundamental Forces exist but there is no reason or means for their existence.
- Constants of Nature have value but do not have a basis in theory for that value to be what it is.
- Virtual particles are added to derived concepts for the purpose of prediction.
- The disciplines of Physics, Chemistry, Biology, and Astronomy do not have a common basis in theory.
- Pure self-isolating, yet successful in prediction, abstract mathematical formulations are used, without known physical counterparts, to describe a physical process in *as if it where* terms. And Physicist, Dr. John Archibald Wheeler of Princeton, may he rest in peace, speaking about the success of Mathematics and Physics.
  - "SOMETHING IS MISSING"

S'energy and The Concept of the Prime Wave

What is needed for a Theory Of Everything as expressed by Kitty Ferguson, in her 1991 biography of Stephen Hawking, eloquently presented his seven "challenges" to the TOE theorist.

Challenges for "Theory Of Everything" Candidates

It must give us a model that unifies the forces and particles.

It must answer the question, What are the "boundary conditions" of the universe, the conditions at the very instant of beginning, before any time whatsoever passed?

It must allow few options. It ought to be "restrictive." It should, for instance, predict precisely how many types of particles there are. If it leaves options, it must somehow account for the fact that we have the universe we have and not a slightly different one.

It should contain few arbitrary elements.

It must predict a universe like the universe we observe or else explain convincingly why there's a discrepancy.

It should be simple, although it must allow for enormous complexity.

S'energy and The Concept of the Prime Wave

- The Physicist John Archibald Wheeler of Princeton wrote:

    Behind it all

    Is surely and idea so simple,

    so beautiful,

    so compelling that when-

    In a decade, a century,

    Or a millennium-

    we grasp it,

    we will all say to each other,

    How could it have been otherwise?

    How could we have been so stupid

    for so long?

It must solve the enigma of combining Einstein's theory of general relativity (a theory we use to explain gravity) with quantum mechanics (the theory we use when talking about the other three forces). (pgs. 20-21)

S'energy and The Concept of the Prime Wave

<u>To this I must add</u>

> *What is the physical mechanism and means of consistent comparable physical change called time?*
>
> *What is the foundation for the most utilized relationship in all the sciences, Pi.*
>
> *What is the origin of physical life?*
>
> *What is the relationship between the physical and the abstract in use and practice of defining the physical*

and

> *A true Theory Of Everything must not only explain why a concept or theory is correct but must also give reason for why an accepted concept or theory is wrong.*

In the next few chapters I will present the Concept of the Prime Wave and then review the challenges above and attempt to give a response that suggests that this concept of the Prime Wave is a true candidate for a development of a TOE.

The idea of trying to understand this wonderful world about us through all of the complex teachings and preaching about the

S'energy and The Concept of the Prime Wave

physical can become such a great task that it would become a laughable thought in the realization that some of the greatest minds of all of the sciences have tried and failed. This task is so daunting that it becomes almost automatic to relinquish authority to academic institutions and those who are charged with the unspoken rule or law to search for, find, and report the truth. Entire libraries have been dedicated to each aspect and discipline of the sciences and yet with all of the knowledge and all of the successes there still remains, questions that can not seem to be answered. We are all involved in that classic of all classic "Black Box" experiments that charges us to describe that "something" that we can not completely see, or feel, or taste, or smell, or hear, called the Universe. We are all involved in this experiment during a time in our history that exposes the latest thoughts and experiments as they happen.

Some of us want to know more and seek to understand and discover what is new. We continue in our search without any intention of being at the forefront, or for that matter even being considered as part of this process of discovery. We are not normally considered as part of the team because we have not the

S'energy and The Concept of the Prime Wave

academic/esoteric training and thus can not be expected to understand the perspective that has been developed over so many years. A perspective so well guarded that to even pose a question you must be invited to ask or you will just be ignored. Let me ask you, the reader, one question before you read much farther. Where will the true Theory Of Everything, a TOE, come from? I believe that no one knows but I can guess that the odds are in favor of a creative mind that might not have been trained to think in a specific and acceptable way.

As I sought to understand what was being taught I found great leaps of faith where needed by all who are to believe in the "scientific method" and the search for truth. Take one example associated with one of the greatest formulas of all of modern science

"$E - MC^2$ " E, is energy and it is equivalent to (=), Mass multiplied by the constant C (the speed of light in a vacuum), squared. But this can also be stated as Energy (without physical counterpart), is " all not otherwise defined", being equivalent to (=), Mass, that which we do not understand, (we know about molecules, and atoms and particles and particles are made of

S'energy and The Concept of the Prime Wave

quarks but we do not know what a quark or a Higgs Boson is.) being multiplied by the measured velocity value of a wave/particle as it passes through an empty volume of space (a vacuum) filled with near infinite quantities of virtual particles, such as the graviton… squared. i.e.; All not otherwise defined is equivalent to that which we do not understand being multiplied by the measured velocity of a wave/particle passing through an empty volume of space filled with near infinite quantities of virtual particles, squared (and this without the consideration of the belief that a particle simply may not exist between two points of an observation!)

Please do not misunderstand; I am in complete awe of the accomplishments that modern science has given to us all through hard work and effort. I have experienced the euphoric eureka moments that must have been experienced by those who realized the value of the fruits their efforts. I simply have an idea that I believe to have value, if only as an analogous representation of what a true TOE would need to be like.

The idea that I have has come to be called the concept of the Prime Wave. (PW)

S'energy and The Concept of the Prime Wave

The PW concept is a description of a physical process in what it is perceived to be terms.

The PW concept has seven, simple to complex, levels of defined interaction that serve to isolate, in definition, a condition, or level, that should not be overly exposed in isolation (as a molecule is separate in level six from the atom at level five.) The seven levels, of the PW, exist as a description of conditions of all that precedes and part of all that proceeds from any point of reference. A certain uncertainty will exist when one condition is exposed and defined from the perspective of another level. The value of this uncertainty will be established as that which is denied. Another condition that inserts uncertainty into the results evolves out of the use of summary reducing equations that serve with great predictive power to achieve generalized or net results in "as if they are" conditions without further reductive support. This can be best expressed in a situation such that at level two of the PW concept, where the Prime Wave is produced, a component of structure is exposed and becomes, in the level four structures, a fundamental condition and this fundamental condition only exists at level four complexities. Under this

S'energy and The Concept of the Prime Wave

circumstance the level two component can be said to not exist in the summary reduced expression of the fundamental condition as perceived from the level four conditions. Thus the Force of Gravity can be defined in *net* terms perceived from level four without consideration of a level two component as with Relativity. The graviton in Quantum Mechanics is a virtual particle and is an expression of the *specific* value of the effect caused by the level two Prime Waves being exchanged by level four structures. Both definitions can be consistent with the definition of each portion from the perspective of each point of view **in isolation**.

In fact the concept of the Prime Wave promotes that all of the Fundamental Forces are as a result of the conditions of interactions of higher than level two structures producing and reacting to the receipt of level two Prime Waves under specific, summary reduced conditions. In addition, all matter, and charge, waves and particles, symmetry, and symmetry breaking, global and local conditions, are all as the result of the exchange of, interaction of, and the existence of, Prime Waves in progression. Thus without pursuing the concept of the Prime Wave farther at

S'energy and The Concept of the Prime Wave

this point, it is best to suggest that a tapestry was woven as a single thread of energy in energy and thus the details of each loop, upon which a pure abstract mathematical formulation can rest with impunity, may be exposed and defined in many ways, i.e.; as a particle or as a wave, as chaotic or stable, as having symmetry or symmetry that is broken.

To the best of my knowledge I have made only one assumption that S'energy exists and have only one major concern that pertains to the development of a possible Theory Of Everything. The concern about this is extremely basic to all of this developing thought. Two conditions must exist in order to define a process. The first is one of the physical event and the second is one of the abstract comparative. In the form of describing the physical universe both conditions are required to be clearly understood on their own. The foundations for the existing cornerstone concepts and theories, that have not been reconciled, were formed out of study, experimentation and intuition and the recognition of the value of specific mathematical formulations. To make such mathematical connections goes beyond my own ability, but such connections

S'energy and The Concept of the Prime Wave

where formed and due to the successes of prediction, descriptions where imposed as to what was believed to be happening. Over the years these imposed definitions became as perceptually valid as the formulations they represent. To consider one to be wrong or incomplete is to consider that both are wrong or incomplete and to do so without a viable alternative would be to commit an idea to logical suicide. "Something is missing" is such a complete way of looking at the problem that it is difficult to even paraphrase it. The "Something that is missing" is believed by me to be a clearly perceived physical process that would exist without interpretation. The concern that I have is that the abstract formulations and the interpretations imposed may not accurately represent the true physical process but does represent an interpretation that is valid within the context of the presentation that may turn out to be analogous. The Physical exists, has presence, and does so without interpretation. The abstract exists, has not presence, and requires interpretation. Both the clearly perceived physical process and the abstract interpretation are required to accurately define the physical. The assumption that I have made, in the

S'energy and The Concept of the Prime Wave

development of this concept of the PW, is that this Energy has presence, is Substantive, in a volume of space that is known as the physical. The physical universe has proven to be experimentally consistent and has the ability to progress from a high concentration towards a lower concentration, or from a high density towards a lower density, in a consistent and predictable manner and I believe that this physical universe is a result of the interactions of S'energy in S'energy. Structure is noted in the existence of the levels of interaction known as molecules, atoms, and particles. These structures exist in a symbiotic relationship in that it may be convenient to isolate and manipulate each separately but if you do so you are establishing a direct component, or condition, of uncertainty, as that which is denied. In the current definitions you find not only attempted isolation in definition but also isolation in the purist of abstract forms as a mathematical symbol. Such symbols can then be easily manipulated with accuracy but always as a condition of an attempt to summary reduce to the lowest or most fundamental level. This summary reduced way of perceiving the physical can fundamentally deny a process that could cause the condition in

S'energy and The Concept of the Prime Wave

the first place. To this I address the concept of gravity as a ***net*** fundamental condition as defined in Relativity and as a ***specific*** fundamental condition as defined in the virtual graviton of Quantum Mechanics. Both conditions can be right and proper and be predictive but only in the framework within which they reside. It now becomes plausible that one or the other or both definitions may be as an analogous representation of the true conditions that make up this fundamental Force/Field condition. This Fundamental Force/Field, called Gravity, may in fact be a Fundamental Condition of the circumstances surrounding the defined effect. That is to say, without the circumstances then the condition may not exist as represented. The classic thought about a tree falling in the wood and the question of does it make a sound if no one is around to hear it acts as a representation of a similar example to that which I speak. Without definition of or knowledge of the compression waves that makes up the sound then it is similar to the problem of gravity in that the tree falls and the ear hears but if there is no ear to hear then it can be said that a net condition is established around a circumstance and that the space around the tree is warped in a specific way and at a

S'energy and The Concept of the Prime Wave

specific time, when the tree falls, such that if you where there to hear it or not the sound exists as a condition of space and time. Summary reduction allows you to completely ignore the process of the wave propagation and view the process as a specific and fundamental condition of Space/Time and through measurements be able to predict the affect at a specific distance over a specific time line. Again when viewed from a different perspective of the ear receiving an event of sound waves or a virtual quantum of energy then this can be further reduced to a specific packet or condition and that this virtual packet or condition exists without a known means. In both perspectives the true understanding of the development of and propagation of the compression waves in air can be ignored and may seem to not be there at all. In fact no process needs to be there for the definitions to be accurate in prediction. Using this overly simplified analogous description of sound and the comparison to the definition of gravity in Space/Time or quantum relationships, I hoped to express an idea that when, at the turn of the century, thoughts developed and described, in very different ways, a single condition called gravity. And that, with all of the success in predicting the affects

S'energy and The Concept of the Prime Wave

of the process, an imposed definition may have caused the disallowance, by definition, of the means by which the condition called gravity could occur and still be expressed, in summary. Another example of this thinking might be that you can predict when the sound from a lightning strike will hit your ear and maybe make the assumption that space is curved for sound and not for light and it takes sound longer to arrive at your ear than light to your eyes? This leads farther to the thought that without a truly accurate perception of the physical process then reconciliation becomes impossible for no one can logically grant a higher value to one perspective over the other until or unless a viable alternative is proposed and Ockham's Razor is applied.

S'energy and The Concept of the Prime Wave

CHAPTER THREE

<u>Building the Universe One Step up at a Time.</u>

My thinking was flawed from the very beginning for I presumed to ask
the "what is it" question and not be satisfied with "what it is like" answers.
And then
I found that S'energy could be "all not otherwise defined".
R. Shriner- 1973 rev:2012

"It turns out to be very difficult to devise a theory to describe the universe all in one go. Instead, we break the problem up into bits and invent a number of partial theories. Each of these partial theories describes and predicts a certain limited class of observations, neglecting the effects of other quantities, or representing them by simple sets of numbers. It may be that this approach is completely wrong. If everything in the universe depends on everything else in a fundamental way"
S. Hawking (1996) pg. 11.

It is my belief that this approach of breaking the work down into smaller parts was not completely wrong in fact it was completely right at the time and in light of the successes, but when you begin to think in terms of a Theory of Everything, a TOE, you must understand that it is quite probable that an intuitive definition, imposed on a successful prediction, may have been wrong. So

S'energy and The Concept of the Prime Wave

now what? Some parts of a concept that is accepted as successful might be right and some parts might be wrong? The problem now is that not only are you looking for a TOE that explains it all but you must look for a TOE that explains why the other concepts do not explain it all! So where to begin? I suggest that before you continue reading this presentation that you take a moment and set aside all that you think you know as a truth. Not to ignore or deny this knowledge but to allow another path to be explored without imposing preconceived ideas too early on the process being described. As a life student I seek to understand what people think about what is happening in the physical but find that for the most part they are always 100% sure they are right and leave no room for additional considerations and re-considerations. They are right and everyone else is wrong...end of discussion. If you are one of these people I wish you all the luck in the world and ask you not to continue the read for you will not understand that this presentation is more a perspective and not a truth without fault. It is difficult enough to think in overview terms and far more difficult to try to explain every detail in your imposed terms that

S'energy and The Concept of the Prime Wave

can never be expanded or changed in any way. As an example, I hope to prove that Time has a foundation in real physical terms. Time is a measure of consistent and comparable physical change. What this means is that changes in the physical world occur constantly and is so consistent that we can observe and build an abstract representation of the passing of time, via a clock, and work with the idea that, the abstraction, can be validly imposed anywhere. Hey it worked for Einstein, when he found no means for such a supportive physical change to occur so he successfully used the abstract of Time to make a prediction and then he denied any physical means for the changes to occur. Is there a physical means for Time, or consistent physical change, to occur? Yes I believe there is a means that allows time to be used in the abstraction while it is understood that in reality physical change is occurring without a clock to measure it. To impose another level of definition on the process of the physical that defines the means for time to exist as a consistent physical change that is comparable is to begin well below any existing level of understanding as presented in the cornerstone concepts and theories that exist in today's definitions of the physical. To

S'energy and The Concept of the Prime Wave

begin this journey of the concept of the Prime Wave you must go even deeper into reality to find a method, a condition, a means for such comparative change to occur at all. It is here that I begin a description of the physical by assuming that the lowest common denominator of all things physical is a dynamic, unique, fluid like substance in a volume of space and is called energy and designated as Substantive Energy or S'energy. What would the Qualities of S'energy have to be to achieve such a complex physical process capable of understanding itself?

We are all within the greatest Black Box experiment of them all. Without an understanding of how it all fits together we roll and twist and bounce and tilt and tap and weigh and listen and measure and suggest and, in the end, hope we will get a good grade.

The journey that I have taken is one that anyone could have taken at any time of day or night in the minds search for understanding of the world of the physical, perhaps even as an analogous understanding, that leaps into the forefront of the mind as a possibility. If it is a good possibility then the mind and body participate as one glob that dances three feet off the ground and

S'energy and The Concept of the Prime Wave

surges with unlimited power, and this is called a Grand Eureka Moment. Have you ever had such a moment, and if not, are you looking for such to experience it for yourself. I was fortunate enough to have had the experience a long time ago and often fondly remember it. I experienced this Grand Eureka Moment but knew that such a profound idea must be proven or ignored. I was trapped for I needed expertise to prove the idea and needed proof to gather the expertise. Forty years ago I explored the Universe in my mind with the belief that I, for the first time, knew how it all fit together....I still find pieces that add to the understanding of the process of the physical so I know it is incomplete but all in all it has been one heck of a journey.

Briefly stated: The concept of the Prime Wave begins here.

"The concept of the Prime Wave is both a perspective and a process that evolves from the belief that the physical exists entirely in a three dimensional volume of space. The physical exists and changes in a consistent and comparable manner and needs not definition. The abstract exists and requires definition. Both a clearly defined physical process and an abstract ability to

S'energy and The Concept of the Prime Wave

predict must be combined to achieve a true Theory Of Everything. The lowest common denominator of all things physical is Substantive Energy, or S'energy. A clearly defined consistent and comparable physical process is the framework that supports and allows for pure, self-isolating, abstract mathematical formulations that are capable of predicting those changes. A single wave form exists of S'energy and in S'energy and this wave, the Prime Wave, is the source of all physical Forces and all physical Structures and is the source of consistent physical change called Time. The Prime Wave origin is a singularity of a maximum density achievement per unit volume. S'energy is the background density as is all convergent Prime Waves incident on a point that can combine to achieve that maximum density value. The Prime Wave expands spherically from its point of origin, has a constant duration associated with the radius of the point of origin area, a constant velocity, and decreases in density inversely as the surface area of the progression increases. Change and Time begins with each new

S'energy and The Concept of the Prime Wave

Prime Wave. The consistency of S'energy to support the Prime Wave in radial progression dictates a condition, a velocity and a distance in progression that supports and becomes the reason and means for Pi. The typical Prime Wave is a condition of the convergence of Prime Wave portions that form nodes that achieve a threshold or maximum density and this New Prime Wave is identical to its parents in every way except where and when and this ability to produce a self-likeness is the foundation for physical Life and a living Universe. This being said there are a total of seven, simple to complex, levels of definition of a single physical process proposed within the concept of the Prime Wave. In comparison with all other cornerstone concepts and theories the concept of the Prime Wave is unique for it builds a physical universe up the ladder to achieve existing summary reductions currently developed from the top down. In the teachings of today it is not uncommon to speak of a Molecule and then an Atom and then a Particle and then a quark and then....? The Structure of a Particle and the condition of

S'energy and The Concept of the Prime Wave

a quark, in the Prime Wave concept, does not exist until Level four with S'energy defined at level one, the Prime Wave defined at level two, the Packet defined at level three, and the Particle at level four. The Particle then exists as a condition of S'energy, the Prime Waves, and the Packets defined in detail first and then the structure of the Particle is identified and is supported by a process of interactions while in summary the conclusion to speak of a Particle in isolation is still supported. (Shriner, 1973 revised 2012)

On completion of the understanding, that I hope to impart to you, you will find that the seven, simple to complex, levels of interaction will explain in such a supportive way as to allow for the mathematics to be predictable in isolation and also allow for a definition of a single process that supports but modifies the description of conditions to include three additional levels of action and interaction below that of the particle:

S'energy and The Concept of the Prime Wave

The Seven levels of interaction are:

1. S'energy

2. The Prime Wave

3. The packet, quantum, photon, or Matrix

4. The Particle

5. The Atom

6. The Molecule

7. The Cosmological

Before you begin to walk this path with me, you must first ask yourself the question, Why would I want to do so? With all of the successes and achievements that have propelled us all into world of understanding, why would you seek to understand or even play with a new view of how this world we live in works? Perhaps it is because we have cornerstone concepts and theories that cannot be wholly reconciled. Perhaps it is a feeling of concern that one of the leading Concepts for a Theory of Everything is couched in poly-dimensional space without known physical counterparts and this leads to the process of the physical being defined in pure abstraction and as a pure abstraction. Perhaps it is because we have a host of Constants of Nature that

S'energy and The Concept of the Prime Wave

have proven values but no basis in theory for that value to be what it is. Perhaps it is the search for answers about Pi, or the Higgs' Boson, or the source or validity of Dark Energy, or Dark Matter, or a reason for Wave/Particle duality, or what a Quark is, complete with all of its color and charm. Perhaps it is just an itch to explore new ideas. Perhaps there are high levels of research and studies that are fragmented and self-isolating in pure abstract mathematical formulations that can predict but defined in such a way as to be very near a curiosity and not a piece to a larger puzzle.

Dr. J. A. Wheeler, formerly of Princeton University, may he rest easy, once spoke of the successes of the mathematics of Physics and commented that with all of the successes still "..something is missing."!

S'energy and The Concept of the Prime Wave

CHAPTER FOUR

Ponder a moment

As a perspective that we all live with, within today's teachings, it is like looking at water and not seeing what details lay within and then ignoring them even when the details are suspected. This would be a vision at the turn of the last century when, looking into the details of matter, no support was found for change, for time, for action, for reaction, wave/particle duality, for Force, for Charge, for fields, for local and global limits, for consistency, for an upper velocity limit, for structure, for mass, for energy values and heat, for Uncertainty, for Constants and for a means for all of the above.

Here are some items to ponder and refresh your mind as to what you believe them to be: These are common references to key subjects so take a moment to consider how they relate to each other as well.

S'energy and The Concept of the Prime Wave

Quantum mechanics

Relativity

String theory

The Standard Model

Constants of Nature

Force Particles

Wave/Particle duality

Uncertainty

Quarks

The Higgs' Boson

The Big Bang

Dark Energy

Dark Matter

Black Holes

Origin of Physical life

Reason and means for Pi

Renormalization

First law of Thermodynamics

Second Law of Thermodynamics

Virtual Particles

S'energy and The Concept of the Prime Wave

All of these and more are exposed as isolated partial descriptions of a single process that underlies the whole of the physical. All of these items are deemed valuable additions to our understanding of the physical in one way or another so let me ask you to consider my belief that for a true physical process definition to merge the thinking into one cohesive relationship it must not only answer questions but should also grant a perspective that supports a reason why they were accepted as being right and why they can no longer be accepted as directly or universally applicable.

Energy is a term used more and more every day and seems to be quite valuable but have you ever considered that it does not exist. This is stated in spite of Sir I. Newton's declaration that Energy can neither be created nor destroyed only modified. Energy does not exist in the physical, it is a calculated condition of potential and kinetic valuations and is a pure abstract term having meaning to some as being "... all not otherwise defined." But what if Energy really does exist in the physical! What would its properties or qualities have to be? Is it passive or dynamic? Is

S'energy and The Concept of the Prime Wave

Energy the method and the means for all things physical? My conclusion is without a doubt, YES! Yes, Energy is the dynamic lowest common denominator for all things physical but it must first exist in a volume of space. Energy in this frame of perspective is more than the potential to do work, it is also the means to do work. Without further delay I present my thoughts of the Concept of the Prime Wave.

$$S'e_{\text{(physical)}} \quad \underline{\textit{is}} \quad E_{\text{(abstract)}}$$

S'energy and The Concept of the Prime Wave

CHAPTER FIVE

Building the Universe One Step up at a Time.

Level one - S'energy

Level One

Substantive Energy or S'energy exists and has three Fundamental Qualities:

- S'energy exists in a three dimensional volume of space and has Density.

- S'energy exists in a three dimensional volume of space and has an Entropic quality that seeks to progress outward from a maximum density achievement.

    (see note #1)

- S'energy exists in a three dimensional volume of space and it has a Viscous Quality that seeks to maintain continuity in and through change.

    (see note #2)

S'energy and The Concept of the Prime Wave

S'energy is not passive. S'energy is dynamic and is the source of all action, reaction, the consistency of change and is also the means for Structure, Forces, and Fields as well as the basis for limits and boundaries of scale and velocity. S'energy exists, cannot be created nor destroyed only modified and S'energy is the means for work to be performed.

> *(note #1: If entropy is the measure of the disorder of a system and in the second law of thermodynamics it is that the entropy of an isolated system increases, then it is quite a problem to have structure, charge, fields, forces and consistency of any kind. Within the concept of the Prime Wave the Quality of Entropy is associated with a fundamental Quality of S'energy and is the single condition of expansion away from a node of maximum density achievement. Entropic expansion away from a source is a fundamental quality of change and is one of the fundamental Qualities of S'energy but it is a specific condition exposed at the achievement of a maximum density and the creation of a Prime Wave. Beyond the level two Prime Wave origins and the harmonics of the*

S'energy and The Concept of the Prime Wave

Prime Wave interactive in level three is the condition where the accepted description is and can be applied. The source condition of entropy is S'energy of level one. The means by which change occurs is by the Prime Wave of level two. The structure that exists through harmonic interactions of Prime Wave convergences is the Packet of level three. The Particle of level four, is a condition of lateral, linear, and opposed directional displacement bonding of level three packets. The Particle structure is in direct opposition to the basics of the second law of thermodynamics but this law is a product of observation of level four and above structures. The perspective is that the second law of thermodynamics applies to levels of interaction not associated with simple to complex interactive of the Prime Wave of S'energy in S'energy and when the law was created no knowledge of the underlying structures existed. Once a structure is created and it continues through harmonic balances then all other conditions not associated with the origin of these structures are subject to a condition of disorder.)

S'energy and The Concept of the Prime Wave

*(**note #2:** If the Viscous Quality, the seeking of continuity, does not exist, then the continuity, the consistency, the comparative, the Rule of Law, the upper and lower boundary limits, and the global and local conditions associated with all pure and predictive perceptions of what is thought to exist, could not.*

If the Viscous Quality, the seeking of continuity, does not exist then the continuity of the individual Prime Wave would be lost, the progression, and density change, would all be random and the Prime Wave would not exist. The ability of S'energy to support this form of Wave progression is the basis for Pi and physical Time, Physical time begins and begins with each new Prime Wave that builds level by level into a consistent and comparable means for change.)

S'energy and The Concept of the Prime Wave

CHAPTER SIX

Building the Universe One Step up at a Time.

Level Two – The Prime Wave

Level Two

The Prime Wave is a compression wave of S'energy in S'energy born out of a maximum density achievement.

Typically this point like, or nodal, maximum density achievement and the resulting Prime Wave, as a single event from a single location, is typically without consequence as the density decreases through progression but this single Prime Wave is the basis for consistent and comparable physical change called Time, the foundation for Fundamental Forces, and the means by which all structures are able to exist through harmonic balance of new threshold achievements, progressions, and new convergences to new threshold achievements. The typical Prime Wave is identical to its parents except where and when and is born out of convergent Prime Waves at a node that achieves the maximum threshold density. The Prime Wave progresses

S'energy and The Concept of the Prime Wave

outward from a maximum density achievement at a velocity ~C the speed of light. (The photon or Packet of level three is an interactive of Prime Waves in convergent sets that has achieved harmonic balance. As the Prime Wave progresses to the core of a Packet and back it must do so while the Photon or Packet is stepping forward toward its own leading edge input at C.) The Prime Wave has a duration of the radius of the volume of space that achieved the threshold maximum density. The Prime Wave progresses outward from its point of origin as an expanding hollow sphere, loosing density directly as the area of the surface of that sphere increases.

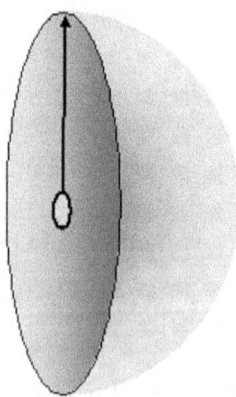

Figure 3: The expanding Prime Wave is spherical.

S'energy and The Concept of the Prime Wave

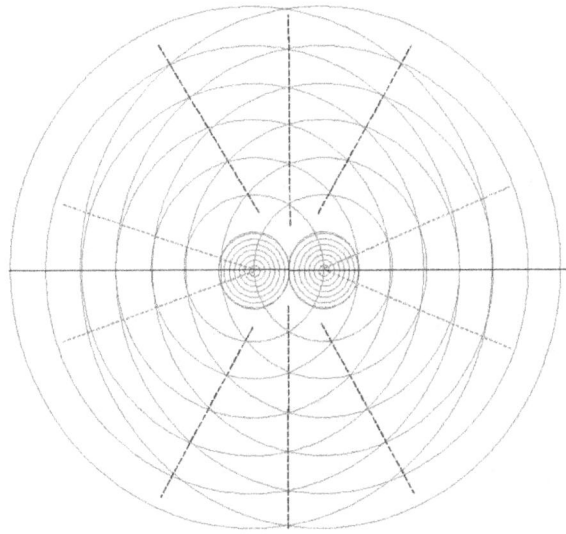

Fig 4 and 5: Two or more Prime Waves converge in two primary ways: The expanding ring of convergence and the cone of convergence. Two Prime Waves are depicted here but when defined as a Packet it is no longer a pair of events not repeated but a set of events that repeat continuously.

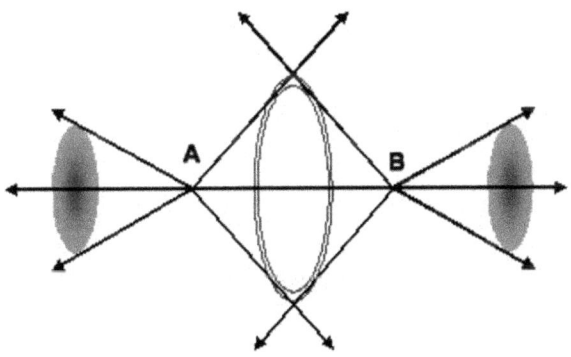

The entropically expanding Prime Wave maintains a viscous continuity through progression and convergences. The direction

S'energy and The Concept of the Prime Wave

of a Prime Wave is always outward and thus time is always forward. A high density achievement is a defined nodal condition that is the source of the Higgs' Boson, the Electro and Magnetic lines of force, the Aural limits, the foundation for harmonic sets and the basis for Structure in the Harmonic sets. This simple Prime Wave stands alone as the tic of the universal clock of change but also is a component of all other conditions known as the physical. Like a mini-Big Bang that is not the largest event but rather the individual Prime Wave is the smallest event of the physical and unlike the Big Bang as being a single event, the creation of Prime Waves occur and re-occur consistently in the harmonic sets of structure.

Seven, simple to complex, levels of definition of the physical process have been identified in definition of a single process called the physical. The simple, as an example, of level two, the Prime Wave, is a single Prime Wave in a volume of space while the complex is the active and interactive structure of a quantum, or packet, or a photon, as the complex of level two. Using this method of summarizing the complex to a simple condition allows for the transition to the simple aspect of level three as the

S'energy and The Concept of the Prime Wave

complex aspect of level two and so on. This definition is for simplification of description of the process only and is imposed on a perceived physical process that at the highest levels still has a foundation in the Qualities of S'energy and of the Prime Wave of S'energy in S'energy. It is consistent to consider the definitions in terms of levels for as we screw down through the levels of complexity as from Molecules to Atoms to Particles we summarize conditions and move on. In this presentation of the Prime Wave I am asking you to take a new look as I screw up the levels of definition and present you with more and more complexity that becomes specific conditions of interaction to be known as Particles, Atoms, and Molecules of levels four, five and six, or Fundamental Force Conditions and in doing so I will comment in notes that, hopefully will merge existing thoughts with this new concept as a condition with two or more descriptions as in an exchange of a graviton or the distortion of Space/Time around a body with mass, or the input of Prime Waves from a consistent source location achieving and affecting a stable structure called a particle (of level four) built with packets (of level three) of Prime Waves (of level two) in sets of

S'energy and The Concept of the Prime Wave threshold achieving Prime Waves in harmonic balance as a throughput engine, a Packet, Quantum, Photon, and a Higgs Particle

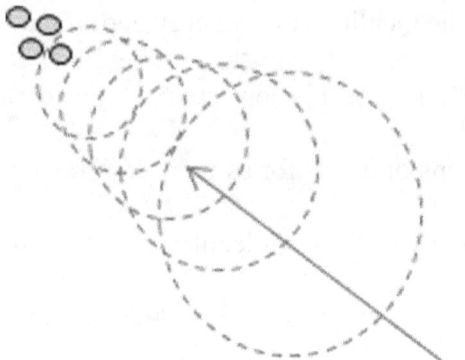

Fig. 6: A single Packet in radiant heat.

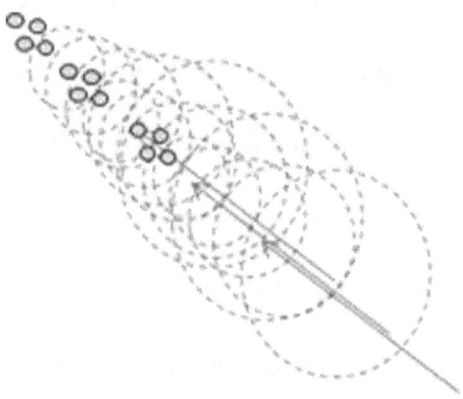

Fig. 7: Linear Bonding of Packets in the Photo Spectrum

S'energy and The Concept of the Prime Wave

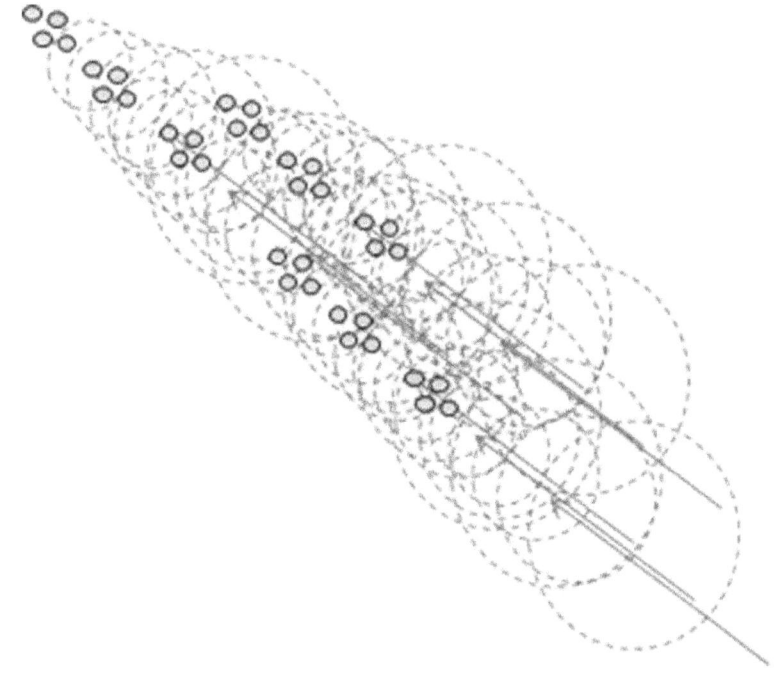

Figure 8
Lateral Bonding of Packets in the Ray Spectrum

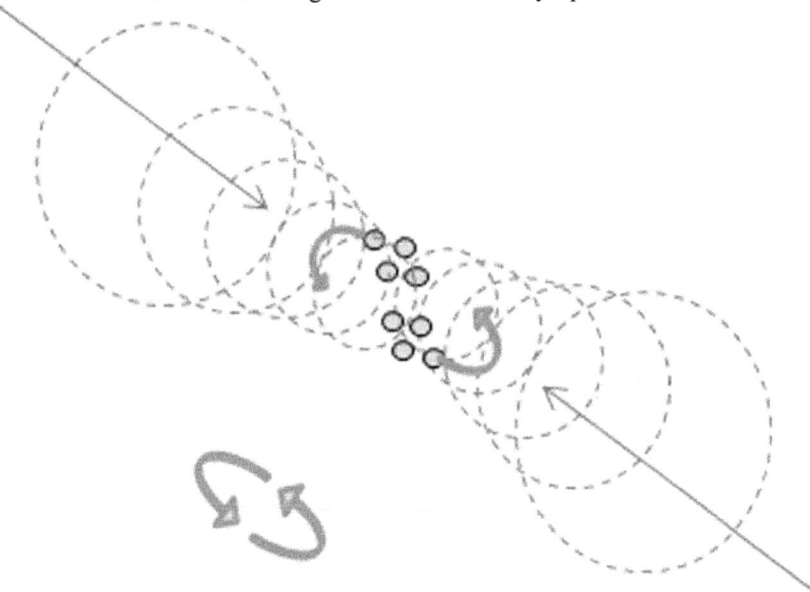

Fig 9: Opposed directional Bonding of Packets in the Particle Spectrum.

S'energy and The Concept of the Prime Wave

S'energy and The Concept of the Prime Wave

CHAPTER SEVEN

Building the Universe One Step up at a Time.

Level Three – The Packet

Level Three

The single process of S'energy in S'energy at a defined level two complex and level three simple interaction. The Quantum, or Packet, or Photon is a throughput engine that does not exist except in the harmonic of threshold achieving nodes and progressions of Prime Waves. (A two dimensional slice through a volume of space would show a background density, Prime Waves being created, progressing, and converging, but no specific structure, force, or field.) The nodes of a packet are equal angular and of a distance that when four or more Prime Waves converge at a node they combine to achieve a new Prime Wave. (see note #3). It is estimated at this juncture that there are 3 sets of 6 nodal components for a stable packet or photon in the volume of space that contains the events (each new Prime wave

S'energy and The Concept of the Prime Wave

is an event) and these events are supportive of the harmonic balance then you have a structure, the first structure of the physical process.

Several conditions must be addressed.

- First there is a background density, consisting of S'energy and all Prime Waves passing through that volume of space.

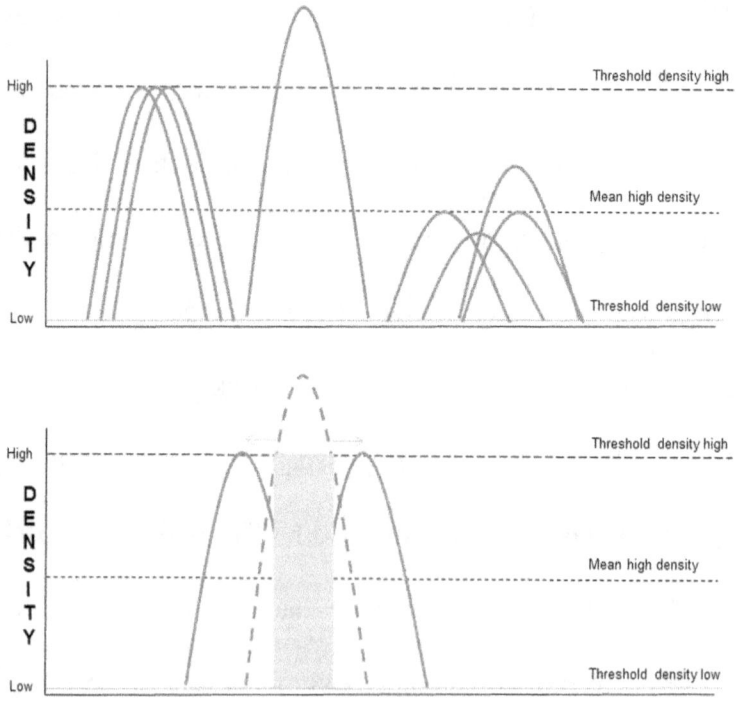

Figures 10 & 11

Limits of background density and the Prime Waves.

S'energy and The Concept of the Prime Wave

S'energy has three Fundamental Qualities and I have focused on the promotion of the Prime Wave with a priori affects of Entropy and Viscosity but it must be made clear that when the Prime Wave, and / or S'energy as a whole, progresses and diminishes then it reaches the Minimum density value whereby the Prime Wave is no more and blends back into the Background density or Heat value of the volume of space. S'energy as a whole expands outward from the stable sources of Prime Waves (like a series of Pings) but also S'energy seeks to maintain continuity as a whole and it is here that a priori affect is changed from the net Entropic expansion to a net Viscous contraction. At level seven of this concept the Universe, the Forces and All structures evolve farther outward to a minimum density value and thus becomes part of the Background density of the whole in a cycle that begins and begins as time begins and begins with each new cycle or Prime Wave and ends when a Prime Wave is no more.

- Second is that there is a maximum density value, a Mean High density value and a minimum density value. The Mean High density value is a condition of the net Heat value of the space under consideration. (see note #4)

S'energy and The Concept of the Prime Wave

- Third is that a high density Prime Wave, that distance beyond achieving the maximum density which created the Prime Wave and while progressing outward but before the individual Prime Wave decreases in value to fall below that of the Mean High density value is a highly influential condition that is able to support additional high density nodes and to cause them to achieve the Maximum density value and form a new Prime Wave. This distance is very small for an individual Prime Wave but in sets, from a specific area location, the convergences extend the area of influence as in a charge field of level four.
- Fourth condition to be addressed is the density of input to the stable Packet. This input is typically from a specific direction, as the Prime Waves are in motion and the convergences to threshold are also, and the net input direction to the Packet is its own leading edge. The Packet is, in the stable harmonic, displaced or stepping forward toward that leading edge input. The velocity of

S'energy and The Concept of the Prime Wave

the displacement of the Packet is known as the Speed of Light.

- Fifth condition is that the displacement of the Packet toward its own leading edge is a virtual straight line but this direction of travel is subject to change in the direction of a new input of comparable value from the side and this will have the effect of changing the direction of the packet. If the input is of low value but consistent as an input direction (as in passing a sun) the influence is minor on a step by step basis and an arc is traced. If the input value is high and the input direction is from one source the change in direction of the Packet as a whole will be very dramatic and would appear to an observer as a reflection or a bounce. If the input direction, from the side, is of high density and from all sides the packet will slow down, as in passing through water, until the input values are passed, then the packet will speed up again to the "speed of light in a vacuum".

- Sixth condition associated with a packet is again the influence of input to the Harmonic structure. As noted

S'energy and The Concept of the Prime Wave

earlier the Packet, in a vacuum, is progressing at the speed of light, this means that lost to the Packet Prime Waves have no repeatable source location and the influence of sets of lost to the Packet Prime Waves is reduced to a very small area and notable in the direction opposite to that of the leading edge or the trailing end of the Packet. If two or more Packets are progressing along the same line or direction and one is in the lead then the second is influenced to follow this high density input that trails off side to side. The second Packet is Bonded in a linear fashion to the first Packet even if the first Packet is influenced by a low density but consistent input. A chain of Packets can thus occur and this chain of Packets become the Photo Spectrum with each link of Packets changing the energy content of the Photon.

- Seventh condition associated with this Packet is that, as with the sixth condition, a Packet can not only have a linear Bond but can also have a lateral displacement Bond as the progressing first Packet and second Packet (and more) influence each other as high density inputs from

S'energy and The Concept of the Prime Wave

the side. Each Packet has its own harmonic structure yet this input synchronized with each other in the harmonic of the sets and their lateral Bond is a displacement that is secondary to the leading edge input. This lateral Bonding of Packets establishes the basis for the Ray Spectrum.

- Eighth condition is the most important for a complex structure called a Particle of level four and it is the opposed directional displacement Bond. Two Packets of similar net values pass each other in opposite directions and close enough to each other such that the high density lost to the Packet Prime Waves mutually change the direction of progression of each Packet in the direction of where the other Packet was. (they chase each others tails) With only two the influence is on a two dimensional plane and is subject to outside input that can break the displacement Bond. If however four or more are opposed directionally Bonded and at least one is not on the same two dimensional plane then, with four, a structure is created that has six identifiable zones and this is the first Particle. (see note #5)

S'energy and The Concept of the Prime Wave

(Note #3: As an example 4 nodes A, B, C, and D achieve threshold density and a new Prime Wave occurs simultaneously at each node. These four new Prime Wave nodes are equal distant and equal angular. The four Prime Waves converge at point E (at the center of this tri-mid) and achieve an additional or 5th Prime Wave. The five Prime Waves created, recon verge at the positions of A, and B, and C, and D and with a new convergent Prime Wave E (plus the background density) produce an equivalent set to begin the process again. This sequence continues to repeat. This example is too small and too ridged but it does serve to represent the harmonic structure mechanism:)

(Note #4: Recall that a Prime Wave decreases in density as it progresses outward from a point of maximum density achievement. The outward progression of the Prime Wave occurs and this high density wave is, for a specific distance associated with the source point location, above the mean high density value and will continue to loose density down through the mean high value, and also continue, with continuity, within and below the mean high value. Although this is a minor point here, it is a condition of defining a limit for the charge field of level four, the Particle. The mean high value is the Heat content of the space that

S'energy and The Concept of the Prime Wave

surrounds the packet as an input density to the harmonic of the packet. This condition changes the effective area of the charge field as a high mean high value decreases the area of influence of the individual Prime Wave. High densities of the Prime Waves and a low mean high value increases the effective area of the charge field. This Heat content or Mean High density value for the space surrounding a structure is always changing locally.)

(Note #5 The first Particle is of four Packets opposed directionally bonded yet still progressing forward toward their own leading edge input at a velocity of C, the speed of light. The simple aspect of level three is the formation of the throughput engine called a Packet. The complex of level three is the Photo Spectrum, the Ray Spectrum, and the Particle Spectrum. This complex aspect of level three is referenced here for it is still S'energy and Prime Waves.) The Prime Wave, as stated earlier, is the basic component of change, consistent, and comparable change called Time. Although there is a great deal of detail left out of this discussion I hope that the fundamental aspect of a physical means for the existence of physical Time to be, is possible and more than that probable.

S'energy and The Concept of the Prime Wave

S'energy and The Concept of the Prime Wave

CHAPTER EIGHT

Building the Universe One Step up at a Time.

Level Four – The Particle

Level Four

The first Particle is the complex of level Three and begins the simple aspect definition of level Four. Earlier it was pointed out that the Particle consists of packets that are opposed directionally displacement Bonded to each other and an example of four nodes was given but for simplicity this same example can be applied to four packets. The three key zones (of six) are the central area, the shell area and the charge field area. The shell area is further defined as that zone where the packets exist and continue to progress toward their own leading edge input and are influenced to achieve a circular path by the High Density input from other Bonded packets in the Particle. Opposed directionally Bonded Packets within the Particle make up the origin of the Particle but as more and more Packets are added to the Particle, you will find

S'energy and The Concept of the Prime Wave

lateral and linear Displacement Bonding also occurring. Some Particles are stable and some are not and most of the reason for non-stable Particles is due to an imbalance of Bonding conditions. At a packet count of Six equal spaced and equal distant from each other you will find the first Quark Pair. Prime Number reductions within the count gives the Up and Down Quarks which are two subsets of three or three subsets of two Packets and these sub sets are created out of lateral and linear Bonding. This stable Particle is the Electron. (5 packet count is the Positron) A Particle of six Packets is always an electron with a specific Charge field shape, a specific dimension, a specific Mass, and this is so no matter where or when or what variables are evident.

> **(Note#6:** The effect of Heat or the Mean High Density value of the space that surrounds the Particle is to extend, or contract the area of influence of the patterns or shape of the charge field (zone three). With an increase in Heat the charge field, the High density area made of Prime Waves lost to the packets bound to a particle that is Zone three, is reduced and a decrease in the Heat will cause the Charge field High Density area to be

S'energy and The Concept of the Prime Wave

extended till the sets of Prime Waves lost become equal to the Mean High Density value that surrounds the Particle.)

Each Packet, Particle Bound, is bound into the particle structure through the exchange of lost to the Packet Prime Waves. These Prime Waves extend beyond each Packet and converge with other Prime Waves also lost to their Packets also bound to the Particle to create the charge field outwardly and the core conditions inwardly. As the Packets are progressing in a circular path, the outflow of Prime Waves form the first spin or spiral condition of all Particles and Atoms. Standing wave nodes of these swept wave convergences form the structure of the Electro and Magnetic lines of Force beyond the charge field limit and establish zone four of a Particle. Zone five of the Particle is the equivalent to the nodes of the charge field but they are not Magnetic as a line of Force but rather are structural points in space and the geometry of these points form the basis for the Aural Force accomplishing lock and key orientations of Atoms, and Molecules in a Biological structure development. Zone six of the Particle is the outflow of the same Prime Waves as before but they have a combined directional influence as an input that is

S'energy and The Concept of the Prime Wave

directional to other Particles, Atoms, and Molecules with mass and this is the basis for the Gravitational Force. There is a limit whereby the individual Prime Waves, even in large quantities but of the same source area, are lost as a directional Force and become part of an expanding outflow over large distances. This outflow also has a limit as the original Fundamental Quality of S'energy comes into play in the seeking to maintain continuity of the whole. This Quality, this Viscous Quality, of S'energy is the basis for the potential condition whereby all S'energy is concentrated in an area and would create the condition known as the Big Bang as the area as a whole achieves a threshold density and begins the process of building the universe up one step at a time.

> **(Note #7**:   As an example of the interaction between two sets of lost to the Particle shell Prime Waves can be envisioned as with two vibrating rods protruding up through the surface of a pan of water. This classic demonstration of waves is also a good representation of a cross-section of sets of lost to the Shell Prime Waves. On an axial line between the two emitters you will find cross convergent nodes. In three dimensions

S'energy and The Concept of the Prime Wave

these are rings of convergence. On this same line beyond each emitter you will find a super positioning pattern. This Super positioning pattern is fully merged at ~<60 deg. In three dimensions this is a conical projection off of the end of both emitters. Each Packet is a throughput engine of sets of Prime Wave occurrences and so the patterns noted above are compounded to include all of the lost to the packet Prime Waves from each to each. Each Packet, bound to a Particle, is subject to creating these patterns of convergence with each other Packet Bound within the Particle shell. For a six unit Particle, the electron, there are 6x5 conical projections each at 60 deg. of projection. The 30 cones x 60 deg. equals 1800 / 360 deg. gives 5 full overlaps. And each projection is beyond each packet so they are offset and the spiral projections are superposed and overlap each other. The offset spiral projections and the superposed projections form Banded areas at specific distances, depending on the geometry of the Shell of the Particle. These banded areas affect other Particle shells and cores to produce a capture condition within the Banded field structure. The high density standing wave nodes of the charge field acts on the core and shell of a Particle to hold the

S'energy and The Concept of the Prime Wave

Particle, as a high density standing wave input, in this high density space or banded pattern as opposed to an area, with no standing wave nodes even though both conditions are created from individual Prime Waves. Also do not forget that each Packet has these same projections as well.)

The geometry of the Bound Packets (in zone two) in a Particle dictates the shape of the charge field (in zone three) and also dictates a new condition within zone one, the core area. As the geometry of the Particle grows with more and more Packets so to does the core area. Within the core area Dependent High Density Nodes are created that achieve Threshold density and establish new Prime Waves but these Prime Waves are not in Harmonic Balance with each other in simple Particles but can become harmonic in Complex Particles, Atoms, and Molecules. They are dependent on the geometry of the Particle and exist only as a condition of the nodes being re-created from the output of the Packets bound to the shell zone. This Higgs' Particle is a throughput dependent condition that is a time delay of throughput as the influence of these new Prime Waves, of High Density, is to resist the changes of the location of each Packet

S'energy and The Concept of the Prime Wave

and thus the whole Particle is influenced to go back to the position or condition it was in when the dependent wave was created. High variations of input density affect the overall position of the Particle and these changes of location of the Particle in a vacuum form the basis for Brownian motion. You have the number of Packets in a specific geometry in the shell with a specific set of dependent nodes in the central core producing Prime Waves again as a throughput engine, and you have a charge field of zone three that is also dependent on the geometry of the shell packets, which gives, when combined, a specific Particle Mass density and charge shape per unit volume. You have a means for the majority of the Constants of Nature.

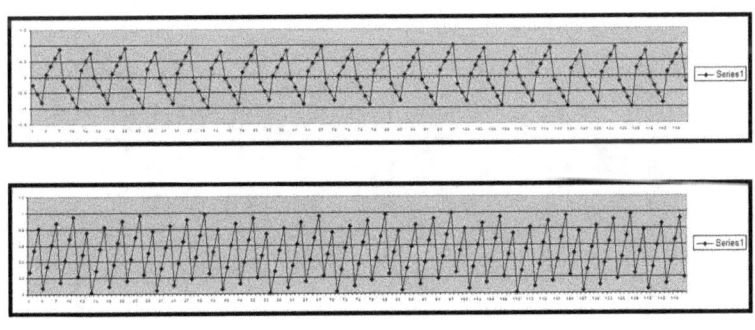

Fig 12 &13: Cone and ring projected surface area patterns –Particles.

S'energy and The Concept of the Prime Wave

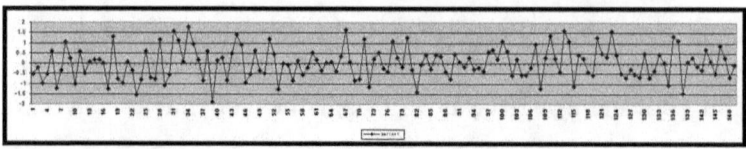

Fig 14: Particle Spectrum is displayed as an overly simplified projection of the cone and ring patterns between Packets bound with each pattern reversing with each addition of 3 Packets as a new overlay is achieved at 180 deg. for each and without modifying the core dimensions as Packets are added. (Note that #6 is – and #5 is + and K Mesons appear at position 12.)

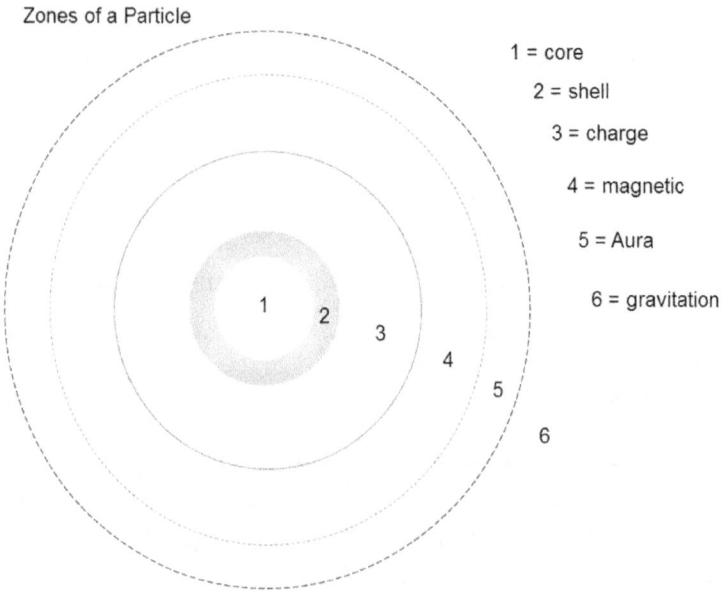

Fig 15: Zones of a Particle.

Zone 1: The core.

The core is an area filled with lost to the Packets of Zone 2 Prime Waves progressing inward. This zone is a net high density area where convergent

S'energy and The Concept of the Prime Wave

Prime Waves may achieve a threshold density that produces new Prime Waves. When harmonic sets of the new Prime Waves are dependently created it is the source of the condition known as the Higgs' Boson or "god" Particle. This condition is a throughput delay mechanism that seeks to return the packets and thus the Particle to where it was. The Particle geometry dictates the number and location of dependent nodes that do achieve the threshold density and so becomes the mechanism of resistance to change of the whole.

Zone 2: The shell.

The shell is a condition whereby Packets are being directionally influenced by lost to the Packets Prime Waves from the second and other Packets mutually. The Packet is a set of Prime Waves being created and re-created in the harmonic. The forward direction of the Packet is a displacement toward its leading edge input, at C, the speed of Light, normally but not in the Particle shell as a great input direction and density value is not from the leading edge but rather from where the other Packets where when they lost their Prime Waves that progressed inward to zone 1 and outward to zone 3 through zone 2.

Zone 3: The Charge Field.

The charge field is a condition of lost to the Packet Prime Waves outward bound from zone 2 and zone 1. There are three forms of convergent Prime Wave Patterns, the individual Prime Waves from separate Packets, the cone and the ring. As the Packets progress at the speed of light around the second

S'energy and The Concept of the Prime Wave

zone the out flow patterns effect any Packet or Particle that enters into the zone. As these patterns are also harmonic and may contain high density nodes as standing wave nodes, the spiral patterns are consistent with the Packet geometry within the shell every time. This swept pattern is the foundation for vortex or circular effects and patterns at all upper levels of definition.

Zone 4: Electro and Magnetic.

The magnetic field is a condition of high density nodes of Prime Waves lost to the Packets in zone 2 and are beyond the limit of the charge zone 3. These high density nodes are standing wave nodes that exist with every Particle but it is not until the geometry of groups of Particles, Atoms, Molecules converge and combine to form swept patterns of the nodes. If the EM nodes are of an elements structure, a stationary magnetic field is created. If the source Particles are in motion then the nodes progress outward as a wave of nodes.

Zone 5: The Aural zone.

The Aural zone is a condition very like the EM nodes of zone 4. The field also consists of nodes of high density convergent Prime Waves as before. The difference is that these nodes are not swept as sets but exist as harmonic sets due to the geometry of the Atoms and Molecules. These nodes are not a magnetic field swept configuration but rather are spatial as a fixed (reoccurring location) pattern of individual nodes. This pattern has the effect of establishing an orientation or Atoms and Molecules in the surrounding area and is the lock and key mechanism for Molecular orientation and selection as in a DNA Molecule. This nodal condition exists at zone one and

S'energy and The Concept of the Prime Wave

becomes the Higgs' Particle if the nodes achieve threshold density. They exist at zone two in each of the Packets of level three as the geometry of the Packet is realized. They exist as nodes in zone four as part of the Electro and Magnetic fields, And they exist in zone five, the aural field, as a spacial set that creates a pattern that is non-magnetic, yet they affect any other Particle or Atom or Molecule in the zone in a way that acts on the zone one Higgs' Particle to influence distance, location, and orientation. To be effective these nodes must have a specific density greater than the mean high density of the space that surrounds. When the mean high density raises the effect of the nodes diminishes and so too does the charge field effect as a distance from source. The Aura is the Lock and Key of the Biological selection of Atoms and Molecules while also being the means for establishing the density value of the states of Matter in Chemistry.

Zone 6: Gravitational.

The zone of Gravity is a condition of a consistent source of Prime Waves being produced. Zones one, three, four and five are high density convergent Prime Waves in progression inward and outward from the Packets in the Shell zone2. These convergent patterns are no longer effectual in zone 6 as high density patterns but rather as an outflow direction of Prime Waves from a consistent source. The mutual output of these Prime Waves from each Particle, Atom and Molecule is directional fro a consistent source and thus as a directional into to all other Particles, Atom, and Molecules. Packets are influenced by input direction and it is normal for a Packet to be displaced toward its own leading edge but in a influence from the side as in passing by

S'energy and The Concept of the Prime Wave

a sun or close to another Particle will affect the direction of the path of the packet. This directional node patterns in zone1, the core, and this directional displacement is in the direction of that input. With multiple and variable inputs to an Atom, the Atom will appear to dance in Brownian motion.

CHAPTER NINE

Building the Universe One Step up at a Time.

Level Five – The Atom

Level Five

The complex of level four is the stable Particle which becomes the simple aspect of level Five.

To be clear, the Particle is a very complex structure with six identifiable zones of interaction. Each zone is a product of throughput. The throughput is specific to the geometry of the Particle as determined by the number of Packets contained. The smaller the number of Packets bound to a Particle the less likely that a Higgs Particle will occur and the less likely the Particle will display conditions of Mass. The more Packets that are bound to a Particle, in the shell zone, the larger the influence of sets of Prime Waves. All of the conditions of all of the zones are specific to the geometry of the Packets bound. All of the conditions occur at an interactive velocity of slightly faster than

S'energy and The Concept of the Prime Wave

the speed of light. (ref. level two above) This complex interactive dynamic is a far cry different than the idea that a Particle is an inert gob of matter. When necessary to give an example, I shall, but for the most part- this presentation is to promote a perspective that builds up from the most fundamental level of S'energy, and progresses through the levels of the Prime Wave, the Packet, the Particle, and now the Atom. As a visualist, I view the processes as a whole and examine the components as needed to expose or promote a condition. A written description of this complex vision is far harder to accomplish than it was for me to recognize and develop the idea in the first place. If I make a mistake or confuse the issue by dwelling on a specific point or condition, I apologize for my inability to explain a process that no one has presented before. If I am wrong so be it, but if I am right our view of the Universe will change forever.

Within the primary Structure of the Particle, three zones (of six) have been identified as the central core, the shell, and the charge field. The stable particles of the Electron, the Proton and the Neutron have been experimentally identified as having a

S'energy and The Concept of the Prime Wave

negative, a positive, and a neutral charge field and a difference in Mass. The structure of the simple Atom of level five is the interaction of each Particle to each other Particles' charge field. Within the concept of the Prime Wave this interaction will be the affects and effects of each zone of a Particle as it relates to each zone of another Particle. The Mass of the Proton (and Neutron) is significantly larger than the electron and so too is the complexity of the geometry of the shells but the charge field is still only a projection of the two forms of Prime Wave convergences that form the charge field. Every Particle has this same combination with the conical and the ring patterns overlay only on different scales of volume and complexity. These projections sweep outward and overlap each other to form banded structures within the charge field zone. Although these charge field structures are of Prime Waves they are superposed upon each other and with the higher density, depending on the specific pattern positioning, one banded area causes a greater affect on another Particle Shell than the other. Thus an electron shell can become completely captured within the charge field pattern of a Proton while the Proton shell is barely affected by the output of the lost to the

S'energy and The Concept of the Prime Wave

Packet, outward bound Prime Waves from the electron. This influences the over all pattern as being an imbalance to the over all charge field pattern projected by both. This imbalance is the foundation for the Ion effect. The Proton and Neutron are far more complex and have much larger charge field patterns and when they are caused to approach closely to each other, the entire particle shell of each Particle is captured in the charge field banding of the other. The shape of the charge field band is dictated by the geometry of the shell patterns and the particles are bound within each other's banded patterns while retaining all other characteristics. It is believed that the Banded Charge field structure of the Proton is a harmonic opposite to the structure of the charge field of the Neutron. This condition of harmonic opposites forms the basis for the *Strong or Nuclear Force*. The only variation of this is that the two patterns are made up of Prime Waves and the charge field of each combine or merge into a new pattern. Super positioning of the patterns in the charge field zone creates the same form of dependent threshold achieving nodes that act in the same way as the Higgs' particle nodes within the central core zone of each Particle. These nodes,

S'energy and The Concept of the Prime Wave

again, are dependent on the geometry of the structures of the Particles involved. If you raise the heat value or the Mean High density of the volume that surrounds the Atom, a very specific condition exists, where by the nodes can become specifically spaced and angular and in the harmonic of the Prime Waves and the Particle structures bound. These sets of nodes can become Harmonically stable packets with one being created right after another, a chain of packets is created and a photon, a set of photons are created of specific frequencies matched to the particles bound. This pattern is, again, specific to the structures bound and is the means for the *Photo-Spectrum of the Elements*. As the Mean High Density value that surrounds the volume is decreased the charge field is expanded as the superposed Prime Waves extend the affect farther in this form of convergent stacking. Complex Atoms have an added factor in the super positioned patterns are standing wave nodes repeated at specific geometric distances and these extensions form the basis for solidification and the means for *States of Matter*. The fourth Zone of a Particle is now a condition to be identified in these standing wave nodes. I call them the Electromagnetic fine

S'energy and The Concept of the Prime Wave

structure nodes that make up the electro and magnetic lines of force of swept convergences resulting from the spin of the Particles caused by the packets progressing at the speed of light in the shell zone of the particle and sweeping the lost to the packet Prime Waves out through the charge field zone but as individual Prime Waves maintaining continuity in progression through convergences but at this distance the individual Prime Waves are of a value below the Mean High density value and is hidden as an individual wave. The nodes make up the form and means of *Electromagnetic Force*. Zone four Electromagnetic fine structure nodes have a structural relationship to each other as a result of the sweeping patterns from each Particle. When these nodes are not in sets an additional condition exists as they exist as individual nodes (higher density than the normal for that location) and these individual nodes are still projected by all Particles and all Atoms and all Molecules but it is not until the complex of level seven that these nodes are recognized for what they are as *an Aura* of a complex Atom or Molecule. It is the Aural field that rotates and dictates the condition of an element of two Atoms or more bound to each other. These standing wave

S'energy and The Concept of the Prime Wave

nodes of the Aural field act as high density inputs that surround and penetrate into zone 1 (the core) of another Particle, an Atom, or a Molecule in the proximity. These nodes become the lock and key mechanism of the Molecular selection, orientation and subsequent locking in to a position dependent on the other structures bound in the Atomic or Molecular structures.

As represented here, each level is constructed from convergences of Prime Waves in progression. Each Level is only a definition of a single process broken down into smaller chunks. A great deal, again, is being stated and unsupported here but the perspective is, again, that a single process can be defined that is three dimensional and able to support a reasoning for successful in prediction concepts and theories that where created by reducing the known down to ever smaller components. The concept of the Prime Wave is an attempted description of the same process from an entirely new direction and perspective, as from the bottom up. A specific point here is that nothing represented here in the form of a Force or Structure is from other than the creation and progression of Prime Waves of S'energy and in S'energy. *All changes that occur and all structures and*

S'energy and The Concept of the Prime Wave

all conditions of Force are as the result of these Prime Waves and this is the form and means for consistent, comparable, change called Time. All Prime Waves are created from the convergences of Prime Wave portions (and the background density) into nodes that achieve the threshold density. All Forces and all Structures are also as a result of these convergences and as all Prime Waves are as a direct result of these convergences they are identical to the parent Prime Waves except where and when and this is the form and means of a *living Universe*.

CHAPTER TEN

Building the Universe One Step up at a Time.

Level Six – The Molecule

Level Six

The complex Atom of level five is the simple of level six, the Molecule.

Level six is by far the most complex of all the levels unless you consider that the simple of each level is as valid as the complex of each level and all the variables in-between make up a single process of progressions and the convergences of Prime Waves of and in S'energy.

The Ionic patterns of charge fields of multiple Particles become simple shadows and diffused patterns of one Particle charge field pattern behind another Particle shell. In fact values of Quantity, Density, Distance, Direction, Duration, Diffusion, and Duration are all factors that must be considered at each level and no less here. Bonding of Atoms into the table of the elements as a

S'energy and The Concept of the Prime Wave

Molecule is a mirror image of one Atom with its duplicate and these charge field convergences are complimentary in the harmonic of equivalent structures. As the dependent nodes of the Aura are projected into the volume of space that surrounds the Atom and the Molecule they may become a magnetic line of force or a string of Packets that make up a photon, they may become the Higgs' Particle within zone one of a particle or they may be the means for a change of orientation of a single molecule within a larger structure and change the makeup to become a fingernail or a nerve cell. The projected conical patterns of convergences from a packet bound to a Particle core are of sixty degrees of angular sweep and as they only exist in pairs of patterns they represent, when combined, 1/3 of 360 deg. of rotation for each pair. The circular patterns of convergences of the cross convergent patterns also being swept outward from a Particle shell and make up 2/3 of the 360 deg. of rotation for each pair. In perfect conditions (which never exist) these patterns are exactly 60 and 120 deg. of patterns projection but in real conditions the angles can be from 57 to 63 deg. or 117 to 126 deg. of projection. This is a condition of offsets of the

S'energy and The Concept of the Prime Wave

subsets of nodes that make up the Packet, the Particle, the Atom and also the Molecule. If you consider that the consistency of the Prime Wave of S'energy and in S'energy and the conditions of progression, density loss, and velocity and you change any one ever so slightly and I would not be writing this. The Prime Wave expands spherically from its point of origin and is part of all that exists to include change. *Pi is not a condition of only one Prime Wave* but is a condition of the structure of the convergences in the Harmonic of Structure beginning with four nodes equal distant and equal angular. Change the Qualities of S'energy and you have a change in the relationship of structure and Pi. ~60 degrees of angle (and harmonics of same) is the more predominant angle of the constructs of Matter and Molecules.

The Molecule exists as a complex structure by itself but there are two conditions of dependence and independence born of the structures and their interactive with other Molecules. The stable Molecule of Elemental Atoms is independent of other structures that may exist in the volume of space that surrounds. The complex of level six is the condition of dependence of the

S'energy and The Concept of the Prime Wave

Molecular structures on each other. As represented in the presentation of the Aural nodes and of the Electromagnetic Fine structure nodes a condition exists whereby the combined structure may or may not require the Aural nodes to exist as a structure. When the Aural nodes are required to sustain a structure you have a dependent condition effecting the structure of changing Molecules, orientation or structure or any combination thereof. This dependence is not to be confused with the projected Charge Field patterns that set up the conditions of *States of Matter* but the Charge Field patterns do play a role in that dependence but of a specific and contiguous nature of the involvement whereas the Aural nodes are not contiguous but are standing wave high density nodes beyond the Charge Field limits.

Comment:

- Form or break a shell zone structure of a Particle and you have the *Weak Force* apparent.
- Form or break a Proton/Neutron Charge Field encapsulation of each to each and you have the *Strong Force* apparent.

S'energy and The Concept of the Prime Wave

- Consistent outputs of sets of Prime Waves lost to a structure beyond the Charge Field limit is the *Gravitational Force* apparent.

- Standing Wave nodes of Prime Waves beyond the Charge Field limit is the *Electromagnetic Line of Force* apparent if the Structure of the source is in alignment and is also the source of the *Aural nodes* apparent that exist but are not magnetic in structure or affect.

- *The Photon is a set of nodes that achieve threshold density* in harmonic sub sets and is both a *Wave* (set of Waves) *and in the harmonic the first Particle*.

- The Structure of a Particle shell dictates it's mass and charge field conditions and is the basis for the *Constants of Nature*.

- *Time, physical time, begins and begins with each new Prime Wave.*

- A summary reduction of all Prime Wave interactions introduces a sphere and a circle into the mechanism of the process of the physical. No other cornerstone concept or theory introduces a means in support of the

S'energy and The Concept of the Prime Wave

use of Pi yet as seen here it is a natural consequence of the interactions of Prime Waves and becomes the *means or origin for the utility of Pi.*

- For a structure to exist in the concept of the Prime Wave a harmonic must occur in the throughput engines of the Packet, the Particle, the Atom, and the Molecule. These harmonic occurrences are resulting from convergences of individual Prime Waves. The New Prime Waves thus created are equal in all respects to its parents except where and when. *This creation and re-creation of Prime Waves is the origin of the physical nature of Nature and life in a living Universe.*

CHAPTER ELEVEN

Building the Universe One Step up at a Time.

Level Seven –The Cosmologic

Level Seven

The complex of level six is the simple of level Seven.

Level seven is unique for it is the only level that builds from the complex to the simple.

The complex of level six could be either a dependent or an independent structure but all exist as a throughput engine of S'energy and in S'energy. From level two to level six, I have presented the condition of the Prime Wave as an origin of physical Time and of the means for the origin of physical life and a living Universe but now we must deal with the other condition of a Prime Wave and all subsets from a given source location in a volume of space and that condition is the depletion of density. S'energy has three Fundamental Qualities being: existence in a volume of space, the Entropic Quality that causes a priori origin

S'energy and The Concept of the Prime Wave

of a Prime Wave as it expands away from and maximum density achievement and the Viscous Quality that is also a priori condition that maintains continuity of the individual Prime Wave but also *seeks to maintain the continuity of S'energy as a whole.* When the Prime Wave becomes a Prime Wave at the point of expansion and the net density falls below the Maximum density value then the Viscous Quality of S'energy plays a role in seeking to maintain the continuity of the individual Prime Wave through progression and convergences. The Viscous Quality is more than a secondary Quality of the Prime Wave for it is the foundation of S'energy maintaining its own integrity as a whole. All space contains S'energy changing of and within itself. The Prime Wave begins a process of expansion that could and does progress outward from a maximum density achievement in a volume of space but the Prime Wave has a finite life and in the progression as an individual wave or as part of a subset or as a condition of a net outflow of waves then this is the thinning of a volume with density. Beyond the limits of the Prime Wave maintaining its own integrity and the combined output of Prime Waves of a given mass and this output falls below the minimum

S'energy and The Concept of the Prime Wave

density value you have the condition where by the Viscous Quality of S'energy takes A priori position and seeks to maintain continuity of S'energy as a Whole. It is the Viscous Quality that seeks to maintain continuity of S'energy as a whole and in the absence of Prime Waves to push the density volume outward, the Universe contracts. On achievement of sets of Prime Waves created in a specific volume of space and in harmonic balance, the process begins and begins but with structure the beginning is not so universal and becomes more localized around that structure as a throughput engine. The Big Bang may have occurred as represented but it is far more probable that a region of space became so dense as to cause Prime Waves to occur individually and as points filling a region. During this moment in a region of space with all of the output/input of Prime Waves, structures are built and maintain their independence through large variations of heat content of the space that surrounds. It is this condition then that produces an even greater output of packets as photons and the creation of stable Particles and these structures are born in an outward flow. These structures, individually and/or as a group, progress outward to a region of

S'energy and The Concept of the Prime Wave

space. As represented in the discussion of the Particles, Atoms, and Molecules an outflow of Prime Waves occur and extend the affect of this output beyond the Charge Field of zone 3 of the Particle and outward beyond the conditions of zones 4, 5, and 6 whereby the influence of the Gravitational Field or Force is no longer a condition of the individual or a combined influence, input directional, as a whole of the output and you have structures like galaxies that are in reality self-isolating yet moving outward as nothing is influencing the whole to stop that outward bound flow. This outward flow will continue until the density of the volume of space, containing the minimum background density value is achieved and the structure depletes and the structure is no more, the source of Prime Waves is no more and the cyclical process begins anew. The Universe is so large that it can be thought of as a dragon so large as to live by eating its own tail when S'energy produces Prime Waves and these Prime Waves progress outward forming Packets, Particles, Atoms, Molecules, and all of the Forces and Fields that surround the structures but decreasing in density until they loose

S'energy and The Concept of the Prime Wave

continuity and merge back into the volume of S'energy as a whole.

Comment:

Dark Energy is the net outbound Prime Waves from an origin but beyond the minimum density value of maintenance of continuity of the individual or sets of Prime Waves from a given source area.

S'energy and The Concept of the Prime Wave

S'energy and The Concept of the Prime Wave

CHAPTER TWELVE

Building the Universe One Step up at a Time.

What if?

What If Newton was wrong and right?
when he denied the integration of Objects, Forces and Laws in order to guarantee the universality of his system and to assert that it would be absurd to believe that gravity was inherent in matter while also promoting the idea that the Objects, the Forces, and the Laws could not be derived one from the other and still was able to predict an outcome?

What If Einstein was wrong and right?
when he denied a physical, tri-dimensional, means for Gravity making it a field rather than a Fundamental Force and, even more, did he ignore, or was he unaware of the existence of, a consistent, comparable, means for the physical process to change and become the ultimate tic of the clock of Time and still was able to predict an outcome?

What If Hawking is wrong and right?
To assert that the partial theories we already have are sufficient to make accurate predictions in all but the most extreme situations, the search for the ultimate theory of the universe seems difficult to justify on practical grounds and still he is able to predict an outcome?

S'energy and The Concept of the Prime Wave

The human mind is quite a fantastic thing. Just short of considering the ultimate Theory Of Everything (because one encompasses the other) is the considerations given to the mind and its abilities. There are two happenings in the mind that needs to be talked about for a few minutes here. We as humans, have the ability to experience our universe and to consider and re-consider those experiences. Born of a physical chain of events that go back in time to the Beginning we have learned to teach that which we have found to be true and then by accepting information as being true, we have a choice to act on those teachings and understandings or not. We have the ability to predict future events by being aware of that which has gone before. In basic terms we know the sun is going to rise over the horizon at some predetermined tic of the clock and we act on that knowledge. We know that if you take a glass of water, in an outstretched hand and let it go that it will fall and make a mess. We know that if you do not eat you will get hungry and being hungry could lead to a much bigger problem if not satisfied. We also have the ability to accept an abstraction from experience and apply it. A track on a muddy path has indications of claws and

S'energy and The Concept of the Prime Wave

the size and depth of the track suggests the weight of the animal that made the track and if it is deep and just filling with water then you can conclude that this imaginary animal is big and has claws and went through the area not too long ago. This imaged animal is determined to be real and is in the local area and because there are older tracks in the same path it comes this way often. You now are faced with two things: to believe your knowledge and experiences and to act on something you can not see or feel or touch and the like. You have made a prediction based entirely on the investigation of the path and find that you need to act and get away or to build a shelter and perhaps a spear and a fire, as night is coming on in one hour and 14 minutes. You have made a prediction based on what you have found and you are acting on that prediction even though you do not know what the animal actually is. You have taken the prediction that the animal is in the area and will return and then you built an abstract scenario about how to protect yourself if you do not leave that path. In a matter of seconds this human mind had an experience, made a prediction based on what had been taught and then built and abstract shelter with a fire and a spear for

S'energy and The Concept of the Prime Wave

protection during the upcoming night. In a future time, you consider what you did and re-consider the facts and acts and determine that this friendly pet on a leash still did not change the potential for harm and next time a bigger spear would be needed. You are unique as a product of an unbroken chain of events going back along a path that encompasses the entire history of the universe. You are unique as a condition of what you have learned and experienced and as a result of the consideration and re-consideration of those experiences and you are aware. You are unique and you are aware and you have the ability to predict. It is this ability to predict out of awareness that allows steps of understanding to grow into actionable truths. You are now on the cusp of taking a pure abstract understanding and building an abstract from an abstract. But what if the truths you where taught and have come to be believed was based on a missed premise, a product of prediction that was made without a full understanding of the parameters that made the first premise viable. Variations of this *building an abstract from an abstract* have occurred over the entire recorded history of man. Empedocles of Acragas 495-435 B.C. embodied the idea of the

S'energy and The Concept of the Prime Wave

nature of the universe by explaining that there are four primary elements, Air, Earth, Fire, and Water held together by Love and Strife. Today we know that this embodiment is reasonable as air, water and earth is a description of the three states of matter as a gas a liquid and as a solid.

Fig 16: Air, Earth, Fire and Water Photos © Jan Shriner

We know that the term fire is the embodiment of chemical activity and combustion and we know that the idea of strife is

S'energy and The Concept of the Prime Wave

embodied in the concept of entropy and the progression toward chaos but we still do not know what the counterpoint to Love is: Gravity, magnetic attraction, Dark Matter or some other exotic condition that seeks to bring things together.

Was Empedocles of Acragas wrong when he built up this understanding 2,500 years ago or was he simply unaware of all of the details of molecules and atoms and particles and fundamental forces and fields that we now know about?

Figure 17: The Ouroboros

S'energy and The Concept of the Prime Wave

The symbol of the Ouroboros represents a dragon eating its own tail, or the universe in a constant state of recurrence. There is no beginning and no end to this cycle with occurrences along the way. Alchemists, during the Renaissance, where very much aware of this basic premise and embraced it. Where they wrong in considering a constantly changing mid-way point along a cyclical path that makes up the universe? Or where they intuitively aware of a process that logically constructs and deconstructs and must build again, somehow?

What if these concepts mentioned above are partially right or partially wrong who would know and would it really make any difference if you believe or if you do not believe.

Building an abstract from an abstract is a great ability in predicting a host of outcomes. Science Fiction writing is a genre of writing that is built entirely on taking a projection to a conclusion and exploring what that projection would be like somewhere in the future.

I have taken the time to explain some thoughts on the ability of the human mind to predict, to build possibilities and to act on

S'energy and The Concept of the Prime Wave

those possibilities and it is here that I must return to the task at hand and the current state of the world of Physics. In a recent article in a magazine the author spoke of the current front runners of concepts seeking to take the lead on a true theory of everything. He sited seven front runners in a profusion of candidates in the battle for the minds of scientists and their search but on the read I was struck by this authors candid remarks of : *despite decades of work, enduring puzzles, precious few clues* and *need not loose too much sleep,* and *if it is ever discovered.* This seeking to understand the *mind of God* is a lofty goal and to speak of an effort toward a given end is far better than to have had the idea and not speak out and present the work believed to have value. Not speaking out is far worse than to have never had the idea in the first place for it would have been a complete waste of time. Seven candidates where presented, each in a candid way, and all of them are abstractions built from experimental fact and abstract premises and projected toward a possible conclusion. Seven foot prints in the muddy path that they believe is going toward a TOE. What a fantastic journey that is represented here in this article as each scientist

S'energy and The Concept of the Prime Wave

applies his mind efforts to achieve an understanding that no one has ever had before. Scientists are on a fantastic journey along a path that they believe will lead them to an understanding of how the universe, and all that is contained within, fits together. Such a fantastic journey but unfortunately, it is my belief, they are all on the wrong path and going in the wrong direction by accepting pure, self –isolating, successful in prediction, mathematical formulations as the final word that, on successful application, requires intuitively inspired definitions to be imposed and applied without a known physical counterpart (String theory takes the lead on this in my opinion). With all of the success this becomes a tough love statement that rests on the idea of building an abstract from an abstract and a single premise that there is no known Tri-dimensional physical means to accomplish all that we experience on a minute by minute basis every day of our lives. (in some cases the means was fully denied – Einstein built an entire theory based on the premise that there is no, tri-dimensional, physical means for consistent, comparable, physical change called Time.) I do not want to know the mind of god. I do not want to build a multiverse. I do want to understand what

S'energy and The Concept of the Prime Wave

is real and what is not. And I believe that Time has a physical means to tic-toc its way forward. As a naive little child I sit on the shoulders of the teachings of some of the greatest minds this world has ever known and ask the question "What if energy is real?"

Comments, Questions and Statements

- I do not know if S'energy is porous.
- Dark Matter is S'energy in a non-structural form.
- String theory is the mathematics of the Harmonic of Prime Wave convergences.
- The Gravitational Force is the exchange of Prime Waves between two stable sources or upon a structure of Prime Wave subsets in harmonic balance as a throughput engine.
- The Graviton is the specific value of the exchange of a Gravitational Force condition. Space/Time is Space containing S'energy acting through Prime Waves on stable structures and from stable structures.

S'energy and The Concept of the Prime Wave

- The Electromagnetic Force is split in the concept of the Prime Wave to separate the structure of a Packet and the condition of a set of standing wave nodes forming ether a line of force or a progressing radio wave. This is specific to the Harmonic structure and the dependence of the nodes. The Higgs Particle is a dependent set of nodes in the core zone of a particle that achieve threshold density and form new Prime Waves, while the EM fine structure, in comparison, have nodes that exist as dependent nodes that normally do not achieve threshold density values. All are nodes under different circumstances of affect and location from the same source condition.

- The concept of the Prime Wave is both a perspective and a means to describe a process called the physical. The physical needs not a definition to exist while the successful pure, self-isolating, abstract mathematical formulations need definition to exist as a part

S'energy and The Concept of the Prime Wave

of an understanding of how it all fits together -

both are required, in my opinion.

S'energy and The Concept of the Prime Wave

CHAPTER THIRTEEN

Building the Universe One Step up at a Time.

"SOMETHING IS MISSING" REVISITED

I have introduced you to the concept of the Prime Wave and have written about a process and a perspective that is uniquely opposed to the current methodology of building an understanding of the physical almost entirely through abstract mathematical formulations that are successful in prediction but seem to impose intuitive notion definitions that are not able to be reconciled. Here are the challenges for a Theory Of Everything as presented by S. Hawking in his book. I will take each challenge, one at a time, and insert the answers to each challenge as developed in the concept of the Prime Wave.

Hearkening back to Ferguson's "Seven Challenges" (1991)....
Challenges for "Theory Of Everything" Candidates

It must give us a model that unifies the forces and particles.

S'energy and The Concept of the Prime Wave

No current theory does unite forces and particles. The C of the PW does build a tri-dimensional process of Prime Waves at level two being created, progressing, and converging to produce a structure at level three, the Packet, that is a throughput engine when stable in harmonic sets of Prime Wave creations. This Packet, bound to each other through displacement bonding in the shell (zone 2) of a Particle of level four then becomes a stable platform for output of Prime Waves and for receiving input from other Particles. This structure at level four that makes up the Particle does so in specific geometric patterns whereby the electron's geometry is stable and reproducible with the same qualities anywhere and any when. As the structure exists in a stable Particle and is producing a stable outflow of Prime Waves then these Prime Waves form specific patterns of convergences that occur with or without another Particle being near or in the specific patterns created. These fields and patterns have specific configurations but always only as a result of the Prime Waves being produced and progressing outward. Each of the zones defined in level four, the Particle, are complimented in convergences with other Particle zones at levels four, five, six

S'energy and The Concept of the Prime Wave

and seven to produce classical Force and zonal effect areas that are specific to the geometries of the structures. The Fundamental Forces are not Fundamental (meaning not able to be further reduced) but they are Force conditions that exist in the C of the PW as a product of events that make up the Particle, the Atom, the Molecule, and the Cosmologic and these lost to the Particle Prime Waves converge and reconverge and produce specific conditions known as Fundamental Forces or Fields.

It must answer the question, What are the "boundary conditions" of the universe, the conditions at the very instant of beginning, before any time whatsoever passed?

No current theory does define the "boundary conditions" of the Universe. The C of the PW does answer these questions at Level one and at Level seven. Level one is the definition of S'energy with fundamental qualities and a defined process progressing upward through each of the levels of definition of structures and forces and fields and then as the progression occurs outward into the cosmologic distances the structures are no longer able to be supported as throughput engines and there is a point where the individual Prime Wave is no longer. The S'energy bound to an

S'energy and The Concept of the Prime Wave

individual Prime Wave structure is lost but the fundamental quality of S'energy called the Viscous quality still seeks to maintain continuity of the whole of S'energy in a volume of space. The Entropic and Viscous qualities of S'energy together form the boundary parameters at both the micro and the macro limits of the universe. The unique function of the Prime Wave being born out of a maximum density achievement and with many Prime Waves being created in a stable structure becomes the foundation for consistent, comparable, physical change called time and is the means for change itself to be and being measured as the tic-toc of a clock so without Prime Waves (level 1) there is no consistent comparable physical change occurring (change yes but consistent and comparable no) This unique perspective of the Prime Wave concept suggests and proposes that the Big Bang may or may not have occurred as a single event but more probably is still occurring as from the origin of each new Prime Wave, again, born out of a maximum density achievement with the differences being where and when.

It must allow few options. It ought to be "restrictive." It should, for instance, predict precisely how many types of

S'energy and The Concept of the Prime Wave

particles there are. If it leaves options, it must somehow account for the fact that we have the universe we have and not a slightly different one.

No current theory sets up the conditions of Constants of Nature or builds a structure that identifies a reason or means for the values to be what they are. By following the C of the PW up through the seven Levels of definition of a single process of tri-dimensional existence it will be noted that there are few options with the exceptions identified as a result of summary reductions of complex events of one level being reduced to a simple aspect of the next higher level. This reduction is necessary in the process of ease of definition and presentation yet the subordinate exists and seems to produce in current theory presentations, without a firm foundation, a measure of Uncertainty as a measure of that which is denied. The universe that we have is a product of S'energy interacting within S'energy itself and any change of these fundamental qualities of S'energy would and could deny the universe that we observe and not a slightly different one.

It should contain few arbitrary elements.

S'energy and The Concept of the Prime Wave

No current theory does establish a complete set of elements that are not arbitrary. The C of the PW does answer these questions at Level one and at Level seven. S'energy exists as the only element of the process of the physical defined.

It must predict a universe like the universe we observe or else explain convincingly why there's a discrepancy.

No current theory does so as successful in prediction, pure abstract, symbols are used to encompass a condition and these symbols, for the most part, have no known physical counterpart. This to include the use of Time as a fourth physical dimension and further to take the lead the concepts of String theory build on eight of eleven abstract dimensions that are, by definition beyond our ability to test. Built within the nature of the C of the PW is a process that is not abstract and arbitrary but mechanical and causal from the origin of a single Prime Wave to the patterns and structures defined along a single path of a defined physical process without denial. From the origin of the single Prime Wave to structure and forces and fields there is a condition of harmonics of the process. The Particle of level four exists as a condition of sets of sets of Prime Waves interacting within a

S'energy and The Concept of the Prime Wave

specific geometry of density and the progression of the Prime Waves outward as an expanding sphere converging and re-converging into new Prime Waves at specific new origin points. Thus a spherically expanding wave form interacting with other spherically expanding waves produce a set of options that are restrictive on the one hand but capable of great diversity on the other. This is best honored on the two part argument about Pi that Pi is a product of Euclidian geometry in three dimensional space, but tell me, if you will, how and why this relationship, this value, is so necessarily utilized in all of the sciences. Pi is perhaps the most utilized value of any single value yet within the existing theories there are no reasons or means given for such value to be derived, used and applied to successful in prediction formulations. A natural consequence of spherically progressing Prime Waves of S'energy and in S'energy and the harmonics of interactions and the causation of consistent, comparable, physical change is the reason and means for the utility of Pi and the support of harmonics from the lowest level of definition to the highest level of complexity.

S'energy and The Concept of the Prime Wave

It should be simple, although it must allow for enormous complexity.

S'energy is energy with all of its potential and both energy and S'energy are "all not currently otherwise defined."

- ***The Physicist John Archibald Wheeler of Princeton writes:***

  Behind it all

  Is surely and idea so simple,

  so beautiful,

  so compelling that when-

  In a decade, a century,

  Or a millennium-

  we grasp it,

  we will all say to each other,

  How could it have been otherwise?

  How could we have been so stupid

  for so long?

S'energy, defined, is what is missing in the development of physical theory and it is, to me, what is behind it all, or rather, what is so <u>simply</u> behind it all.

S'energy and The Concept of the Prime Wave

It must solve the enigma of combining Einstein's theory of general relativity (a theory we use to explain gravity) with quantum mechanics (the theory we use when talking about the other three forces).

No current theory or set of theories are able to suggest even a direction to be applied in the search for reconciliation of the two great theories. Building naturally within the process of the C of the PW is a set of conditions that did not exist in the understanding of the interactive of the physical as the physical process definition was at best denied. (The source of time and the physical means for gravity). Exposed in the body of this description of the concept of the Prime Wave is a process that has seven, simple to complex, levels of interaction defined. The C of the PW builds up the ladder of complexity from level one and the definition of S'energy and in so doing sets up conditions that have not been applied in any way before. Not the passive Ether of old the element of S'energy is dynamic and participates in every aspect of existence. By explanation of the enigma is the understanding that the levels one, two, three, and four did not exist some 100 years ago. As each theory of Relativity and

S'energy and The Concept of the Prime Wave

Quantum Mechanics evolved, into a foundation of our understanding of the physical, they did so without knowledge of a more fundamental set of conditions that is believed to exist now but not then. As is natural in the evolution of the C of the PW, details exist but are summarized into conditions and symbols that allow for predictive maneuvering without having to be immersed in the details of the events that occur. Thus the concept of time is summarized in the abstract term time without the understanding that a physical means exists to support the consistency of physical change. Without a known physical means for consistent physical change to occur, definitions evolved that, in some respects, even denied that a physical process exists to support and cause the changes to occur. Looking down the levels of interaction and not being aware of levels one, two, and three you see only a summary reduced condition or Fundamental condition that, for all practical reasons, needed to be defined. As explained in the body of this text, Gravity is not a Fundamental Force, nor is it a Field, but rather it is a Fundamental Condition of the exchange of Prime Waves progressing outward from stable source locations. Deny that the

S'energy and The Concept of the Prime Wave

Prime Waves exist and you have a field of effect and you can quantify the effect you need while not knowing the what it is that is occurring. Both of these cornerstone theories deny, by their own definition a means, a process, a condition where by something is produced, progresses, is received, and a reaction occurs. They both rely on the success of predicting an outcome either as a **net** macro condition over macro distances or as a **specific** condition of the exchange at micro distances. Both of these theories deny a more fundamental process that underlay the physical process being described and produces the same summary reduced symbols to calculate from. Relativity denies levels one, two, three, and level four interactions as viewed from level four and above, while Quantum Mechanics denies level one, two, and sometimes level three as viewed from level four and above. Both of these theories have intuitively inspired definitions imposed on the physical process portions that have been denied and they are wrong to have done so and certainly to continue to do so. No one can deny the value of the successes of these theories yet they have done quite well without identifying what that *something that is missing* is.

S'energy and The Concept of the Prime Wave

To these challenges I must add:

What is the physical mechanism and means of consistent comparable physical change called time?

No current theory does define a means for Time. The concept of the Prime Wave defines a physical process that, through the creation and progression of the Prime Wave of S'energy and in S'energy and the development of structures, forces, and fields resulting from the interactions of these Prime Waves in progression Time can be both an abstract symbol and a physical process.

What is the origin of physical life?

No current theory does define a means for physical life. At level two the Prime Wave is defined but the overwhelming source of Prime Waves is the stable Particle of level four and this stable Particle is the result of the harmonics of Prime Waves being created, progressing, and converging to achieve a new Prime Wave at a new location equal in all respects to its parents except where and when in a living universe of interactions.

What is the relationship between the physical and the abstract in use and practice of defining the physical.

S'energy and The Concept of the Prime Wave

The terms Energy and Time have no physical counterparts in today's cornerstone concepts and theories. The C of the PW establishes as a basic and fundamental premise, stated or not, that there is a means for both the abstract and the physical to exist and therefore need not to be denied as being real in a three-dimensional universe. Which came first: physical change that is consistent and comparable or the idea that energy and time exist without physical means? What can I say?

and

A true Theory Of Everything must not only explain why a concept or theory is correct but must also give reason for why a concept or theory is wrong.

No current theory approaches an answer to this question while this entire book is filled with proposed answers to this question. Are there questions left sitting on the table, certainly, there are questions such as: is S'energy porous and how did S'energy come to be, and to what extent dependent and independent conditions affect the definition of the process of the C of the PW. How predictable is the zone 5 high density nodes of convergent Prime Waves that make up the 5th fundamental force or field

S'energy and The Concept of the Prime Wave

condition of the Aura. Are there predictions that come out of the testing of the C of the PW such as: Can the Prime Waves that form the gravitational field in zone 6 be blocked or modified to limit the weight of a object on earth? Can a magnetic field in zone 4 be blocked or modified to limit the effects of magnetism on an on demand basis? Can light be focused to produce a field with six passing beams to create electrons as stable particles out of the energy packets at low temperatures and threshold values. Can computer chemistry be performed at a greater accuracy or performance value than experimental? Can non-intrusive diagnostics be performed by reading the aura of a biological system? Can we begin to merge the disciplines of Physics, Chemistry, Biology, Astronomy, and a host of engineering programs into a common thread of understanding? Can we keep the notion of Time Travel? Can we begin to identify and recognize the best way to make or produce a product or process before we build it? What values are natural and understood to be unique as a product of an unbroken chain of events from the beginning of the Universe as we know it?

S'energy and The Concept of the Prime Wave

CHAPTER FOURTEEN

The minds eye.

Man has pursued the understanding of the universe, screwing down into ever deeper depths and ever smaller units, to find the elements at one level, the atom at another, the particle at another, and now we know that a still smaller structure exists within the particle defined within the realm of the quark. The quark, first developed as a mathematical relationship that seemed to describe the changes in the structure of the particle's property's became a reality when a single jet of some unexplained "something" emerged as one particle was smashed into another.

Another level of structure? Another condition of understanding to be explored! Where does it stop? This ever turning screw must surely soon reach the limit of our ability to penetrate, observe and thus to leave only the minds eye to see that which can not be reached otherwise.

S'energy and The Concept of the Prime Wave

The minds' eye seems to have only a self-imposed limit formed by understanding. Once proposed and understood, a concept can expose the thoughts of a structure, in ever-larger scales. So as Einstein rode the winds of change on a photon so to can we ride the winds of change on models that serve to grant insight to levels never before understood or witnessed by man.

It is believed that a single model might be constructed that could resolve many heretofore not reconciled problems of what could be called analogous representations of portions of the physical universe. At least there are four cornerstone concepts and theories that profess to describe what a portion of the universe would be like when viewed from within a specific set of self-isolating parameters.

Energy does not exist in the physical; it is a catch all term used to define "all not otherwise defined". Virtual particles "wink" in and out of existence. An absolute vacuum contains near infinite quantities of these virtual particles as one-definition leaks over into another. We have come to accept these anomalous conditions as a limit to understanding with an understood "Uncertainty". There are well over one hundred "Constants of

S'energy and The Concept of the Prime Wave

Nature" whose repeatable values serve without a basis in theory, while virtual components are added to formulations, without known physical counterparts, in order for the results of the calculation to predict.

Mathematical relationships have been found to be able to predict but definitions may have been imposed upon them in such a way that to question the one is to question the man. Who in his right mind would jeopardize a carrier on a single contrarian thought that requires proof from within the community that holds the definition to be true?

I sometimes think of myself as the child that yells out "the king has no clothes" when I seek help from the spinners of the golden cloth. Spinners who know how to gain favor and obtain the grants and fellowships needed to survive and who also know that to work from within on that which is safe to expand upon is how you secure a future in this world of virtual truths. " Don't rock the boat or you'll be thrown overboard" is a phrase that can not apply for myself, for it is only my wish to understand that leads me down this path and perhaps with a little luck I may pose the

S'energy and The Concept of the Prime Wave

question that could lead to a better understanding of this "wonderful world about"*

What if energy where substantive? What qualities would substantive energy have so as to cause each fundamental condition to exist? And what, if not Substantive energy (S'energy), could exist that would form a foundation for the understanding of the structures of Forces and matter? These questions yet unformed worried me to find a means and an understanding that supported that which had been proven to be true while giving strength to that which is not understood and to bind relationships not reconciled before. The analogous representations formed, in my in my minds eye as I studied, merged one day in a grand fashion as I realized that what I was thinking reconciled, defined, supported, and built upon all that had preceded and could become part of all that may proceed from here. On that day a single understanding was forged before and what was understood as fundamental begins four levels below that of the particle, five levels below the level of the atom, and six levels below that of the molecule. On that day a single thought emerged that allowed energy to be substantive and to

S'energy and The Concept of the Prime Wave

support that understanding by defining three fundamental qualities that serve to produce all of the conditions of Force and Matter as level upon level of interaction formed into level upon level of known structural development. Respectfully I submit that each of the Fundamental Forces of Nature are better described as Fundamental Conditions of the interaction of S'energy in S'energy. The origin of gravity is from level two but the condition of change associated with gravity occurs with the structures of levels three and above. The Origin of the electro and magnetic fields again have an origin at level two but it is not observed as a macro condition until level three and above structures are there to influence and be influenced. The origin of the Strong and the Weak forces again is from level two yet the condition defined does not exist until the structures of the particle and the atom are in place to be acted upon. The structure of the particle has its origin at level three when level two prime waves are such that a harmonic interaction or delayed throughput supports a Packet form of structure that in turn interacts through displacements to form a mutual displacement bonding (there are three forms of displacement bonding, lateral, unidirectional, and

S'energy and The Concept of the Prime Wave

opposed directional) that can become a structure with prime number factors of sets (when disrupted can become the observed quark-ish jet) that establish consistent patterns of structure complete with charge and mass and to exhibit conditions that are defined as the Higgs particle.

Each of the seven defined levels are as a result of those conditions that precede and become part of those conditions that proceed from any point of reference. Be it structure or force or condition all are of S'energy in S'energy. S'energy can not be created or destroyed but the conditions can change. The Prime Wave of level two is the fundamental structure of S'energy in S'energy and is the source of change. This source of change, the Prime Wave of S'energy in S'energy, is the physical side of the definition of time. When the definition of Space/Time was developed no notion of lower levels of structure existed and such the definition served to predict yet the definition imposed denies this mechanical means of directional displacement towards a source. This Prime Wave definition exposes a means for a net displacement in the direction of input while also creating a means for the specific condition of displacement known as the

S'energy and The Concept of the Prime Wave

graviton. The Concept of the Prime Wave establishes a condition of threshold achievement (a maximum density condition) where upon a new point center (a zero point) condition is established of S'energy and in S'energy that forms an electro and magnetic fine structure nodal component that not only sets up the condition of the standing magnetic field but is also the mechanism of the electromagnetic wave and the aura that surrounds a living cell. This EM fine structure node forms the basis for the Higgs particle within the core of the particle and the basis for two forms of charge field surrounding a particle. The nodes, upon achieving threshold density values (much the same as the perceived big bang) can't be established in harmonic sets that act as a stable matrix of achievements that as a packet is displaced forward towards its own leading edge at a maximum rate know as the speed of light. The Packet or photon or quantum of (stable) energy is a self-sustaining structure of the EM fine structural nodes of the EM wave field but the nodes achieve threshold densities in harmonic sets of sets that are of the same condition but of a different density value. The throughput of the packet as it is displaced forward towards its own leading

S'energy and The Concept of the Prime Wave

edge is such that the lost portions of the Prime Waves in the output side are as random waves with out a common source point location, This Lost to the Packet Prime Wave portion forms the basis for the Strong, the Weak, the Charge Field, the Gravitational Field, the Electro and Magnetic Fields, the basis for the Photo Spectrum of the Elements as photons are created within the charge field patterns specific for the particle and atom. The interactions of the Prime Waves form the basis for change and for structure and of conditions of the interaction of one structure to another. Again it must be said that all conditions of Force and of Structure are as conditions of Prime Waves interacting with each other and that all conditions are as a result of levels of interaction defined and imposed. Each condition is a result of all that precedes and is part of all that proceeds from any point of reference.

S'energy and the Prime Wave, the mechanism of change in the Physical, exist without definition while the Abstract requires definition to exist. Both are required for a complete understanding.

S'energy and The Concept of the Prime Wave

CHAPTER FIFTEEN

Building the Universe One Step up at a Time.

Simple:

All that is needed is to understand Substantive energy, or S'energy, is that it is dynamic, exists in a tri-dimensional volume of space, exhibits the three fundamental qualities of density, Entropy and Viscosity in producing a single primary wave form, the Prime Wave of and within S'energy, that when created out of a threshold density achievement begins a process of change which is universally fundamental, directional, consistent, comparable, and is the tic-toc of the universal clock of Time.

The Prime Wave is the smallest single event in the physical process (not the biggest Bang but rather the littlest Ping but very similar in result). Convergent Prime Waves produce new Prime Waves equal in all respects to its' parents except where and when. Sets of reoccurring Prime Wave origins produce a throughput engine called a Packet. Packets step forward, in the

S'energy and The Concept of the Prime Wave

harmonic, toward their own leading edge input at a velocity of c. Packets interact through emitted Prime Waves affecting the direction of travel of other individual Packets, when in close proximity, and when this is opposed directional (LOL two dogs chasing each other's tails) for two or more Packets then a Particle shell is created. Six Particle zones are established in the interactive of the Packets. Prime Waves, emitted in the Particle configuration, produce dependent high density nodes within the core that creates a time delay of throughput known as the Higgs' Particle, these same Prime Waves produce a swept pattern for a charge field, and a set of high density nodes of an electro and magnetic line of force and, these same Prime Waves, create an Aural field again of high density nodes (non-magnetic) in the space beyond the charge field and before the Prime Waves become a directional input to other Particles known as the Gravitational field.

Hypothetically speaking if you were to stop all change and slice a Packet with a two dimensional knife you would not see a packet but only new Prime Waves being created at a high density node, progressing Prime Waves, and Prime Waves merging into

S'energy and The Concept of the Prime Wave

a new high density nodes. The Packet, therefore, only exists in the harmonic of interactions of Prime Waves. The Packet then is both a set of Prime Waves and a Particle or structure at the same time. The Packet steps forward, toward its own leading edge input as a result of sets of Prime Wave convergences, at the velocity of c, the speed of light. As the Packet steps forward at the speed of light the Prime Waves must progress at a velocity just slightly faster than the Packet. Linear sets of Packets make up the Visible Spectrum of Light. Laterally bound Packets make up the Ray Spectrum, and opposed directionally bound Packets make up the Particle Spectrum. The geometry of the Packets bound to a Particle Shell is specific and forms the foundation for Constants of Nature, as each Particle of six Packets bound is always an Electron with all of the specific conditions of charge and mass. Quarks are prime number sub-sets of the packets bound to a Particle shell and these sets are bound to each other laterally as well as through opposed directional displacement bonding. Packets bound to a Particle shell continue to progress forward at the speed of light causing lost to the Packet Prime

S'energy and The Concept of the Prime Wave

Waves to sweep outward from source in two forms, the ring and the conical, establishing two reactive conditions of a charge field. Let's talk about it, OK..

Ok let's talk about any theory or concept that addresses the possibility of a tri-dimensional means for consistent, comparable, change called time.

None?

Ok let's talk about any theory or concept that addresses the possibility of a tri-dimensional means for the equivalence principal to hold true when viewed from afar.

None?

Ok let's talk about any theory or concept that addresses the possibility of a tri-dimensional means for the Uncertainty Principle to hold true when viewed from nearby.

None?

Ok let's talk about any theory or concept that addresses the possibility of a tri-dimensional means for Charge Field to exhibit two different conditions.

None?

S'energy and The Concept of the Prime Wave

Ok let's talk about any theory or concept that addresses the possibility of a tri-dimensional means to reconcile existing cornerstone concepts and theories.

None?

Ok let's talk about any theory or concept that addresses the possibility of a tri-dimensional means for String theory to have value.

None?

Ok let's talk about any theory or concept that addresses the possibility of a tri-dimensional means for the origin of, process through, and aftermath of the Big Bang.

None?

Ok let's talk about any theory or concept that addresses the possibility of a tri-dimensional means for the Higgs particle or "god" particle to be and to effect conditions of Mass, of inertia, and of an object in motion to stay in motion without an outside force acting upon it.

None?

S'energy and The Concept of the Prime Wave

Ok let's talk about any theory or concept that addresses the possibility of a tri-dimensional means for Constants of Nature to have the values that they have.

None?

Ok let's talk about any theory or concept that addresses the possibility of a tri-dimensional means for the origin of physical life.

None?

Ok let's talk about any theory or concept that addresses the possibility of a tri-dimensional means for virtual particles.

None?

Ok let's talk about any theory or concept that addresses the possibility of a tri-dimensional means and support for Physics, and Chemistry, and Biology, and Astronomy, and a host of Engineering disciplines with continuity in development of definition and application.

None?

Ok let's talk about any theory or concept that addresses the possibility of a tri-dimensional means for Global, Local, and Velocity limitations.

S'energy and The Concept of the Prime Wave

None?

Ok let's talk about any theory or concept that addresses the possibility of a tri-dimensional means for action, progression, and reaction at a distance.

None?

Ok let's talk about any theory or concept that addresses the possibility of a tri-dimensional means for light to slow down as it passes through a media and speeds up again on exodus.

None?

Ok let's talk about any theory or concept that addresses the possibility of a tri-dimensional means that supports the pure abstract, successful in prediction, mathematical formulations that are symbolic of a process that is undefined.

None?

Ok let's talk about any theory or concept that addresses the possibility of a tri-dimensional means for the physical.

None?

What do we have to talk about? The concept of the Prime Wave…Ok let's talk about a possible tri-dimensional means for

S'energy and The Concept of the Prime Wave

the physical process to exist today without options of the multiverse and virtual, as if they are, particles.

Let's talk about a living universe.

But wait that is a view that is from outside the box and has been ignored…until now!

S'energy and The Concept of the Prime Wave

CHAPTER SIXTEEN

Building the Universe One Step up at a Time.

Pointing the way to reconciliation!

Modern Physics has stalled, and in some cases individual leaders have even given up, on the search for an ultimate Theory Of Everything.

S. Hawking is one such leader who seems to have come to the conclusion that the concepts and theories that we now have is enough.

I think this attitude by Hawking is wrong to teach. Pure, abstract, self-isolating, successful in prediction, mathematical formulations have come to represent the Physical World and this representation has a problem, to quote J. A. Wheeler,

"...something is missing."

The *something that is missing* is, in my point of view, a clear representation of the Physical Process that underlies and supports the conditions of consistent comparable physical

S'energy and The Concept of the Prime Wave

change called time, the mechanism that supports and establishes the reason and means for an upper limit to the speed of light, that underlies and supports the geometry of interaction to form the foundation for the Constants of Nature, that underlies and supports the conditions known as Fundamental Forces of Nature, and that underlies and supports the basic precept of hidden details proposed in the development of String Theory.

I believe it was K. Quigg, in an article in Scientific American who wrote the definition of Energy as a pure abstract meaning "all not otherwise defined." Energy, as defined by I. Newton, is the potential to do work and cannot be created nor destroyed. This simplified statement has two perspectives as with one having potential without definition of the what, why, how, when and where, and of something that may do something and the second part is of the existence of Energy that is not able to be created nor destroyed. Does Energy exist as a real physical, dynamic, substantive, superfluid that permeates all space and matter or is it enough to deal with Energy as a pure abstract with out physical

S'energy and The Concept of the Prime Wave

counterpart? I. Newton identified, through logic, a potential physical entity that would underlie all physical processes even though he did not or was unable to give meaning to the fundamental qualities that such a substantive component to the physical would have.

The excitement, the successes, the ability to predict, have all engulfed the advancements of modern Physics but have they gone too far into the abstractions of the mathematics of the day and forgotten that a real physical world exists and supports their adventures by being so consistent and comparable from one physical unit that changes to another.

The Question is this "Does the abstract term energy have a physical counterpart and if so what would its fundamental qualities have to be to support all things physical?"

I have found no argument either for or against the existence of a dynamic Substantive Energy, or S'energy. It would appear that a search in this direction is null since the defeat of the concept of the Aether and the advent of the abstract of Time as the fourth physical dimension. Mickelson and

S'energy and The Concept of the Prime Wave

Morley, in their experiment that failed, addressed the idea that the speed of a photon would change as the photon progresses through the passive Aether as the photon progresses into or across the flow of the Aetherial wind. Conclusion, no change in measured speed of the photon and thus no Aetherial wind and thus no Aether. But the Aether was considered to be passive and not dynamic at a most fundamental level and so no further research was needed, except for confirmation. The medium of the Aether would have to be ultimately dense and absolutely thin to support the idea of being a carrier for the wave called a photon that traveled at the speed of light and be infinitely thin as a medium that does not create a headwind as the earth the sun the galaxy passes through it and the universe. The Aether became a dead issue. But wait, what if the Aether was the first blush of a definition of a dynamic substantive energy and the definition failed to give reason or means for such a failed experiment. S'energy is dynamic and not passive as the Aether was defined. To proceed, it becomes a matter of a

S'energy and The Concept of the Prime Wave

definition of the fundamental qualities of S'energy and not a matter of existence and no one is looking.

It is normal for a mathematician to take, or be given, a data set and work out the details of a formulation that would allow the next bit of data to be predicted with a certain amount of success. For over a hundred years details have been experimentally collected and more recently with the advent of computers the amount of details has grown, and grown and grown and with the details comes progress as those details have become more predictable as new ways to view the detailed relationships are formulated and applied. As the success of prediction gave rise to more and more detail the abstract formulations became more and more important and with all of this success came definitions of those abstract formulations. Definitions where generated out of the inspired insight of the creator of the formulation. Definitions where applied, in the abstract, for they did not know then and we still do not know, how the physical process works to give us all that we are able to predict. Again *something is missing* and that *something* is a clear understanding of the physical

S'energy and The Concept of the Prime Wave

process that underlies the conditions that we are able to predict.

Energy is a term used to define the potential, not the means, for work. Time is used as a comparative of the consistency of physical change without regard to the means for that change. Constants of Nature are values, experimentally derived, to be used without reason or means for that value to be what it is. Fundamental Forces of Nature are fundamental because there is no known reason or means for the conditions to exist, but they do. The details of Particles, Charges, and Fields have been found and have produced the valuations and established limits and conditions but the reason, the means, the geometry of that dynamic existence is still unknown. The emphasis on String Theory and poly-dimensional space is an acknowledgement of perceived details and conditions that have no basis in the current list of Cornerstone Concepts and Theories. S'energy, on the other hand, is dynamic and is the perceived means for consistent comparable change called time, is the means for the geometry and interactive of structure, force and fields and is the foundation for a clear

S'energy and The Concept of the Prime Wave

definition of a physical process whose details are exposed in and with the purity of successful mathematical formulations. In a certain reality sense the definition of a physical process building up from the most fundamental level, of seven simple to complex levels, of a defined process there is no longer a problem of reconciliation of the cornerstone concepts and theories as they were developed over time and from the top down and exposed as conditions were found and explored and explained but explained without a common foundation. In other words the condition of a structure such as a Quark is the end product of exploring from the top down while in the building up from the most fundamental level the geometry of interaction of S'energy in S'energy, produces a structure called a Quark that only exists at level four of seven for it is here that the conditions exist to produce such a structure. Without exception or denial the process of defined interactions of S'energy in S'energy support the existence of the conditions defined by each of the cornerstone concepts and theories and more, much more. It is here that a greater understanding is developed in going beyond current Physics

S'energy and The Concept of the Prime Wave

and beyond the reconciliation of existing theories and concepts where there is also the ability to reconcile Physics, Chemistry, Biology and Cosmology as having a single common ground to build up from. Physics is level one, two, and three, Chemistry is all of the below plus level four and five, Biology is all of the below and defined and defended in level six, and Cosmology is the end note of level seven with a return to level one and the foundation of the physical process defined. And to be clear…this reconciliation exists with all of the Engineering and technical nuances in between. The definition of Time begins at level two, the origin of Pi also begins at the simple aspect of level two and the origin of life, in a living Universe, begins at the complex aspect of level two as well. Until now the idea of level one, two and three, as part of seven simple to complex levels of definition of a single process did not exist. Such a foundation for the physical process is not even discussed for it goes beyond the accepted definitions of existing terms bound to pure abstractions formulated to predict and allowed to have no known physical counterpart.

S'energy and The Concept of the Prime Wave

As an outsider I am tired of being ignored and of knocking on the doors and windows of the *Monastic Society* of the world of physics to simply ask a question and have a dialogue about the existence of a dynamic S'energy. I write this book after forty years of effort in the hopes that someone someday may pick up these bits of thought and explore them to some beneficial end.

Robert D. Shriner

"And so it seems to me to be as I now see the thing I think I see"...Again.

Anonymous.

S'energy and The Concept of the Prime Wave

S'energy and The Concept of the Prime Wave

IN CONCLUSION

A Theory Of Everything derived from only one presumptive assumption that substantive energy exists.

A dynamic Substantive Energy, with three fundamental qualities, exists in an undefined volume of space: it cannot be created nor destroyed only modified; it is the <u>means</u> to the potential to do work; It is the <u>means</u> for physical change to occur in a directional, consistent, and comparative manner called Time; It is the lowest common denominator for all Structure and Force calculations; and it is self-limiting.

The specific key to understanding how S'energy becomes the foundation for change and the basis for structure is not as a single event called the Big Bang but rather as the single smallest event, the Prime Wave, that occurs and re-occurs in harmonic structures. Events that act to produce stable throughput engines and the resulting lost to the structure Prime Waves. Each new Prime Wave is the basis for consistent, directional, comparable, change called Time.

Newton deduced rules and laws of physics without knowing the physical process that establishes those rules and laws.

Einstein imposed a description of a physical process that was predictive and did not need a three dimensional physical process to be defined.

Hawking has given up looking for a Theory Of Everything that would include a definition of a physical process in preference to adding bits and pieces to the existing predictive models.

Pure, abstract, self-isolating, successful in prediction, mathematical formulations, spread across seven and more cornerstone concepts and theories, have become accepted as a description of what the

S'energy and The Concept of the Prime Wave

physical is <u>like</u> and ignores or denies any attempt at defining what the physical process <u>is.</u>

Wheeler spoke of the successes of mathematics in physic yet "...*something is missing!*".

The *something that is missing* is, to me, a clear understanding of the physical process that supports the consistency of change, the means to the potential, and produces the global and local limits that are observed. The physical exists without definition while the abstract requires definition to exist. It is here, in the definition of the successful abstract mathematical formulations, definitions brilliantly yet intuitively derived and applied, became imposed for lack of an alternative, a defined physical process capable of supporting the predictions.

A true Theory Of Everything must begin before Time, transition through the beginning of Time, and follow through to the end of Time. It must transition through Physics, Chemistry, Biology, Newtonian mechanics, Astronomy, and Cosmology, and It must be a simple idea capable of great complexity, and through all of this, be consistent.

A Theory Of Everything is a description of the physical process and this process is believed to be Substantive energy, or S'energy, defined and exposed.

$$S'energy_{(physical)} \; \underline{IS} \; energy_{(abstract)}$$

I have an idea that I believe has value. This one idea is captured in

the writing of the book, "S'energy and the Concept of the Prime Wave." It is my belief that failing to express an idea, believed to have value, is far worse than never having had the idea in the first place for it becomes a complete waste of Time. It is my hope that others may read and understand the perspective applied here and take what I have stated to the next higher level.

S'energy and The Concept of the Prime Wave

How far can we go once we understand the mechanics of such a physical process?

There are no practical limits only egos!

S'energy and The Concept of the Prime Wave

REFERENCES

Barnet, L (1948) *The Universe and Dr. Einstein.* New York: Bantam Books. 112-113

Ferguson, K. (1991*). Stephen Hawking Quest for Theory of Everything: The Story of his Life and Work.* New York: Bantam Books. 20-22

Hawking, S. (1996). *A Brief History of Time.* New York: Bantam Books 11-12.

Quigg, K (unknown) Scientific American article.

Wheeler, J. (unknown) speech, "...something is missing!"

Wheeler, J. "Behind it all" unpublished poem published by: Ferguson, K. (1991*). Stephen Hawking Quest for Theory of Everything: The Story of his Life and Work.* New York: Bantam Books. 21

S'energy and The Concept of the Prime Wave

S'energy and The Concept of the Prime Wave

Table of Figures

Fig. #1	Forward	page 15	Abstract Vs. Physical.
Fig. #2	Chapt. 2	page 22	Current Physics 2012.
Fig. #3	Chapt. 6	Page 59	Prime Wave is spherical.
Fig. #4	Chapt. 6	Page 60	Two or more Prime Waves.
Fig. #5	Chapt. 6	Page 60	Two or more Prime Waves.
Fig. #6	Chapt. 6	Page 63	Single Packet.
Fig. #7	Chapt. 6	Page 63	Linear Bonding of Packets.
Fig. #8	Chapt. 6	Page 64	Lateral Bonding of Packets.
Fig. #9	Chapt. 6	Page 64	Opposed directional Bonds.
Fig. #10	Chapt. 7	Page 67	Background density limits.
Fig. #11	Chapt. 7	Page 67	Background density limits.
Fig. #12	Chapt. 8	Page 82	Cone and Rings.
Fig. #13	Chapt. 8	Page 82	Cone and Ring Patterns.
Fig. #14	Chapt. 8	Page 83	Particle Spectrum.
Fig. #15	Chapt. 8	Page 83	Zones of a Particle.
Fig. #16	Chapt. 12	Page 112	Air, Earth, Fire, Water.
Fig. #17	Chapt. 12	Page 113	The Ouroboros.

S'energy and The Concept of the Prime Wave

www.ingramcontent.com/pod-product-compliance
Lightning Source LLC
Chambersburg PA
CBHW051520170526
45165CB00002B/550